Biological and Medical Physics, Biomedical Engineering

For further volumes:
http://www.springer.com/series/3740

The fields of biological and medical physics and biomedical engineering are broad, multidisciplinaryand dynamic. They lie at the crossroads of frontier research in physics, biology, chemistry, andmedicine. The Biological and Medical Physics, Biomedical Engineering Series is intended to be comprehensive, covering a broad range of topics important to the study of the physical, chemical and biological sciences. Its goal is to provide scientists and engineers with textbooks, monographs, and reference works to address the growing need for information.

Books in the series emphasize established and emergent areas of science including molecular, membrane, and mathematical biophysics; photosynthetic energy harvesting and conversion; information processing; physical principles of genetics; sensory communications; automata networks, neural networks, and cellular automata. Equally important will be coverage of applied aspects of biological and medical physics and biomedical engineering such as molecular electronic components and devices, biosensors, medicine, imaging, physical principles of renewable energy production, advanced prostheses, and environmental control and engineering.

Himanshu K. Patel

The Electronic Nose: Artificial Olfaction Technology

Springer

Himanshu K. Patel
Electrical Engineering
Institute of Technology
Nirma University
Ahmedabad, Gujarat
India

ISSN 1618-7210
ISBN 978-81-322-1547-9 ISBN 978-81-322-1548-6 (eBook)
DOI 10.1007/978-81-322-1548-6
Springer New Delhi Heidelberg New York Dordrecht London

Library of Congress Control Number: 2013943329

Printed on acid-free paper

Springer is part of Springer Science+Business Media (www.springer.com)

To

The Almighty Lord Shiva

And

My father Prof. K. J. Patel

"You have given me the power to believe in myself and pursue my dreams. I could never have done this without the faith I have in you"

Foreword by Dr. Anup Singh

Over the last decade, the "electronic sensing" technologies have the fastest and the important developments from a technical and commercial point of view. Usually, it is assumed that Electronic noses (EN/E-nose) are used to monitor complex mixtures of gases. Commercially till date, the E-nose are too expensive and difficult to operate. But in the coming years, researchers plan to develop an E-nose which is easy to operate and robust instrument for vast potential applications.

Many important recent studies have investigated the sensorial analysis of human breath to potentially provide rapid and reliable diagnoses of many human diseases based on the detection of bioindicator compounds and specific mixtures of peculiar VOCs in human breath samples. Prof. Himanshu K. Patel has brought out various aspects of E-nose that includes its real time applications, particularly in human disease diagnosis using exhaled breath, in this compact book. The contents of the book cover most of the requirement of the undergraduate and graduate curriculum. Of course, each topic can be the subject of a detailed study. A student specializing in various subjects will conversely find a common minimum amount of learning in this book. It can be a worthy textbook for the appropriate course in biomedical engineering and instrumentation engineering and reference book for others.

Prof. H. K. Patel's presentation is clear and lucid. His writing style is simple and communicative. He makes a complex subject easy and enjoyable. I compliment him for his utmost effort and his contribution to the "electronic sensing" era. I am sure that the book will be well accepted and will a great success in its area.

Dr. Anup K. Singh
Director General
Nirma University, Ahmedabad, Gujarat, India

Foreword by Shri. Narendra Modiji

MESSAGE

The smell of freshly cooked tea or the hint of a perfume in the air can evoke strong feelings and memories. Apart from our five senses, the human life is dominated by the visual sense but often our deeper emotional responses trigger by smells. Truly our much of the behavior and ecology is ruled by olfactory sensation.

Prof. H. K. Patel with his vast industrial experience and his academic eminence has brought out various aspects of electronic nose in a compact book. I am sure that the book will be welcomed by the student community and become a success in its area.

(Narendra Modi)
Dt. 16-05-2013

To,
Prof. H. K. Patel,
Associate Professor,
Instrumentation & Control Engineering Department,
Institute of Technology,
Nirma University, Ahmedabad.
e-mail : hkpatel@nirmauni.ac.in

Narendra Modi
Chief Minister, Gujarat State

Preface

Time comes when for some conditions breath tests can be performed in lieu of blood samples.

If you want to know all about Electronic nose (EN/E-nose) than this book is for you. The objective of this book is to start with the introduction of measurement technology and then explore the E-nose and its development with applications. It is clear that the scope and importance of making the replica of the mammal system will continue to expand during the twenty-first century.

This book is an introduction to E-nose and aims to ease the tasks of reader who is coming first into get in touch with E-nose. The book is planned to be read on a chapter by chapter basis for those new to the subject and in this format would be appropriated for those on degree, including postgraduate, courses. The text would also be suitable for any reader familiar with the devices but requiring information that takes them somewhat deeper into the detail and applications. For such readers some of the chapters could be excluded and particular chapters studied in more depth. Enthusiastic engineers could find the text helpful as an aid to the development of prototype systems prior to full-scale commercial application.

Chapter 1 is a good starting point; however, it serves as a general introduction to measurement technology.

Chapter 2 presents an overview of the basic principle, characteristics of sensor, and discusses sensor technology.

Chapter 3 is basically an introductory overview of biological olfaction system, signal transduction system, and olfactory pathway in mammalian system. It will help to understand the replica of olfactory system: an E-nose.

Chapter 4—The oldest neural system is the mammalian nervous system. The molecules called odorants is detected by the chemical sense of smell. Chapter begins with the introduction of odor, molecule biology of smell, and various types of odor and reviews its analysis techniques.

Chapter 5—An instrument which would inspect samples of odorous air and report the intensity and quality of an odor without the involvement of a human nose is an artificial nose. This chapter discusses the machine olfaction technology and compares it with mammalian olfaction system. The process of sensation and perception analysis involved in E-nose is described along with overview of different signal analysis techniques.

Chapter 6 provides detailed descriptions of the different types of sensors used in E-nose. This chapter focuses on classification, description, and analysis of most common E-nose sensors. Also a brief idea of recently developed smart sensor using VLSI and nanotechnology is presented.

Chapter 7—Analog, discrete, and digital signals are the raw material of signal processing and analysis. Enough material and books exist for signal processing and analysis. This chapter gives introduction of the signal processing, the conventional architecture of signal processing systems and the notions of analog, discrete, and digital signals are discussed. The final stage in E-nose is pattern analysis. Various pattern analysis techniques are discussed and reviewed in this chapter.

Chapter 8 reviews the widespread application area of the E-nose. The detailed discussion of two case studies presented in this chapter certainly proves the worth of E-nose technology in the coming era.

I hope that this book will be a valuable learning and reference tool for many. I have tried to present basics of E-nose operations and its applications specifically related to biomedical; however, with the unlimited variations in current development of E-nose sensors and systems, I certainly have not been able to provide all details. Only you, prepared with the information gained through this book, can explore the true limits and real-time applications of E-nose.

In the spirit of continuous improvement I am interested in receiving feedback from students, faculty, and professional who use this book.

Himanshu K. Patel

Acknowledgments

This book is a result of my long speculation in the area of Instrumentation and Control. However, it is impossible to neglect all those powerful forces that were always with me during this venture. They not only provided me with a strength (both in the form of motivation and cooperation) which is prerequisite for such work but also made me realize that I am bound to accomplish this difficult task successfully. Even the difficulties and hardships teamed up to strengthen my resolve to complete this book.

First of all, my sincere gratitude to Dr. M. D. Desai for his indefatigable enthusiasm for the study and constructive suggestions that contributed immensely to enrich the book and make it worthwhile.

I am highly grateful to the Institute of Technology, Nirma University for providing all kind of resources that could help me in writing this book. My special thanks to Dr. K. Kotecha for his constant motivation and support. Thanks are also due to Shri K. K. Patel and Shri Ashish Desai for their encouragement. I also wish to thank all the faculty friends and staff members of the Institute of Technology for their invaluable support.

Those we love don't go away; they walk beside us every day.... Unseen, unheard, but always near; still loved, still missed and very dear.

The blessings of my (late) father instilled in me the strength to work hard for the book. I owe to my (late) friend Jay, for his utmost friendship and inspiration. I owe to my family members—My mother, brother, and sister for the shower of love, affection, and support whenever I was in dire need. How can I forget the support of my wife and children who were kept isolated due to my involvement in the book. They always motivated me for completing the task in time.

I am also thankful to my students for their unconditional support whenever required. They were a source of huge inspiration for me in writing this book.

Last but not the least, I am indebted to Shri Narendra Modiji, Honorable Chief Minister of Gujarat State and Dr. Anup Singh, Director General, Nirma University for sparing their precious time and blessing this book with their forward messages. I am truly thankful to Springer for showing confidence in me and giving opportunity to share the knowledge on Electronic nose with readers across the globe.

Himanshu K. Patel

Acknowledgements

Contents

Abbreviation

ANN	Artificial neural network
CNS	Central nervous system
CA	Cluster analysis
cAMP	Cyclic adenosine monophosphate
CP	Conducting polymer
COPD	Chronic obstructive pulmonary disease
EN/E-nose	Electronic nose
EEG	Electroencephalogram
ECG	Electrocardiogram
EPSP	Excitatory post synaptic potential
FET	Field effect transistor
FENO	Fractional exhaled NO
GPCR	G-protein-coupled receptor
GC-MS	Gas chromatography-mass spectrometry
GC	Gas chromatography
HPLC	High pressure liquid chromatography
IR	Infrared
KNN K	Nearest neighbor
LDA	Linear discriminant analysis
MEMS	Micro-electro-mechanical-system
MLP	Multilayer perceptron
MOS	Metal oxide semiconductor
MOSFET	Metal oxide semiconductor field effect transistor
MS	Mass spectrometry
OU	Odor unit
OB	Olfactory bulb
ODT	Odor detection threshold. ODT and OU are used in the U.S.
OFC	The orbitofrontal cortex
OR	Olfactory receptor
ORN	Olfactory receptor neurons
OSN	Olfactory sensory neuron
PARC	Pattern recognition
PC	Principle component
PCA	Principle component analysis

Ppb	Parts per billion
Ppm	Parts per million
QCM	Quartz crystal microbalance
QMB	Quartz microbalance
SOM	Self-organizing map
SAW	Surface acoustic wave
VAP	Ventilator associated pneumonia
VOC	Volatile organic compound

Chapter 1
Measurement Technology

1.1 Introduction

Measurement is the act or the result of a quantitative comparison between a predetermined standard and an unknown magnitude. Scientific and engineering sensors and instrument systems of a spectacular variety of size, weight, cost, complexity, and technology are used in the modern industry. However, a close look would reveal that all of them are composed of a set of typical functional elements connected in a specified way to provide signal in a form necessary. Little progress is possible in any field of investigations without the ability to measure. The progress of measurement is in fact the progress of science and indirectly that of mankind.

> I often say that when you can measure what you are speaking about, and express it in numbers, you know something about it; but when you cannot measure it, when you cannot express it in numbers, your knowledge is of a meager and unsatisfactory kind.
> Lord Kelvin, Electrical Units of Measurement (1883)

Measurement is a vast field which embraces detection, acquisition, control, and analysis of data. It involves the measurement of physical, electrical, mechanical, optical, and chemical quantities and plays a very significant role in every branch of scientific research and engineering process which include control systems, process instrumentation, and data reduction. Measurement is generally specified as a number followed by a unit. All measurements have three apparatus: the estimation, an error bound, and a probability which the actual magnitude lies within the error bound of the estimate [1].

1.2 Significance of Measurement

Measurement is a common string that runs through the fabric of all science and engineering. Measurement is important link in the chain of events in research and development. It begins with a definition of the problem and objectives, and ends with the utilization of information.

H. K. Patel, *The Electronic Nose: Artificial Olfaction Technology*,
Biological and Medical Physics, Biomedical Engineering,
DOI: 10.1007/978-81-322-1548-6_1,

Measurement is necessary for proper operation, maintenance and control of equipment and processes in manufacturing. Without the way for measurement, automation would not be feasible.

In general, there are two major functions of all branches of engineering:

1. Design of equipment and processes and
2. Proper operation, control, and maintenance of processes.

Both these functions require measurements. It is through measurement that a product can be designed or a process be operated with maximum efficiency, minimum cost, and the desired degree of maintainability and reliability.

1.3 Measuring Systems

Measurement is generally made with instruments. Human sense can be very keen but are lacking in range. Measurement can be viewed as the reconstruction of the input from the observed output of an instrument. The input is measurand and the output is the signal from the instrument. A measurement is considered to gather the specifications for accuracy, since a perfect measurement is not possible. Measurements have been involved into every field of scientific activity from medicine to aerospace. Figure 1.1 shows the functional states of measurement systems.

1. *The physical medium* refers to the object where a physical phenomenon is taking place.
2. *The sensor* is affected by the occurrence in the physical medium either through direct or physical contact or through indirect interaction of the observable fact in the medium with several components of the sensing element. The first element in any measuring system is the primary *sensor*: this gives an output that is a function of the measurand (the input applied to it). Most commonly but not for all sensors, this function is at least just about near to linear. Some examples of primary sensors are a liquid-in-glass thermometer, a thermocouple, and a strain gage. In the case of the mercury-in-glass thermometer, the output reading is given in terms of the level of the mercury, and so this particular primary sensor is also a complete measurement system in itself. In other sensors, a

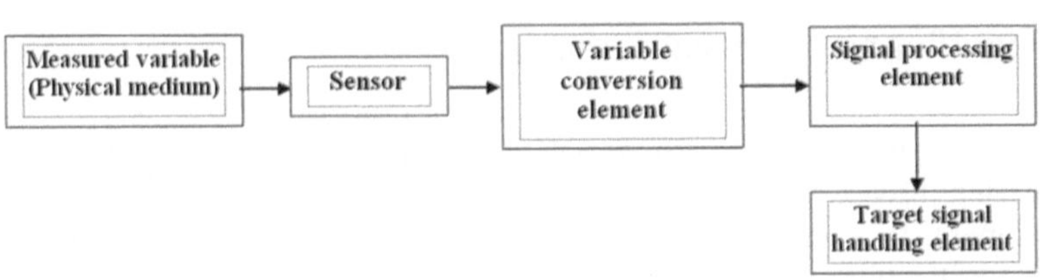

Fig. 1.1 Functional states of measurement systems

signal is directly generated in the sensing element, as is the case of a thermocouple that generates a voltage in response to a difference in temperature between its two ends. Table 1.1 shows the examples of sensor elements.

3. *The variable conversion element* serves the function of altering the nature of the signal generated by the sensor. The method of converting the nature of the signal generated in the sensor to another suitable signal form (usually electrical) depends essentially on the sensor. Individual signal conditioning modules are characteristics of a group of sensing elements. Analog electronic circuits finally produce electrical signals in the form of voltage or current in specific ranges. Variable conversion elements are needed where the output variable of a primary transducer is in an inconvenient form and has to be converted into a more convenient form. For instance, the displacement-measuring strain gage has an output in the form of a varying resistance. The resistance change cannot be easily measured and so it is converted to a change in voltage by a *bridge circuit*, which is a typical example of a variable conversion element. In some cases, the primary sensor and variable conversion element are combined, and the combination is known as a *transducer.*
4. *The signal processing/transmission element* is used to process the signal generated by the first stage for a variety of purposes such as, filtering (to remove noise), diagnostics (to assess the healthiness of the sensor), linearization (to obtain an output which is linearly related with the physical measurand) etc. Signal processing systems are therefore usually more general purpose in nature. Signal transmission is needed when the observation or application point of the output of a measurement system is some distance away from the site of the primary transducer.
5. *The target signal-handling element* may perform a variety of functions depending on the target application. It may therefore contain data/signal display modules, recording or storage modules, or simply a feedback to a process control system. The output of the primary sensing element may be electrical signal of any form. It may be voltage, a frequency or some other electrical parameters. When the elements of an instrument are actually physically separated, it becomes necessary to transmit data from one to another.

Examples contain a temperature chart recorder, an instrumentation tape recorder, a digital display or an analog to digital converter (ADC) tracked by an interface to a process control computer.

1.4 Units of Measurement

Near the beginning systems of measurement were based on whatever was available as a measuring unit. For purposes of measuring length, the human upper body was a convenient tool, and gave us units of the hand, the foot, and the cubit. Although generally adequate for barter trade systems, such measurement units are of course

Table 1.1 Typical examples of sensor elements

S. no.	Input variable to transducer	Output variable of transducer	Principle of operation	Type of device (sensor)
1	Temperature	Voltage	An e.m.f is generated across the junctions of two dissimilar metals or semiconductors when that junction is heated	Thermocouple
2	Temperature	Displacement	There is a thermal expansion in volume when the temperature of liquids or liquid metals is raised and this expansion can be shown as displacement of the liquid in the capillary	Liquid in glass thermometer
3	Temperature	Resistance change	Resistance of pure metal wire with positive temperature coefficient varies with temperature	Resistance thermometer
4	Temperature	Pressure	The pressure of a gas or vapor varies with the change in temperature	Pressure thermometer
5	Displacement	Inductance change	The differential voltage of the two secondary windings varies linearly with the displacement of the magnetic core	Linear variable differential transducer (LVDT)
6	Displacement	Resistance change	Positioning of a slider varies the resistance in a potentiometer or a bridge circuit	Potentiometric Device
7	Motion	Voltage	Relative motion of a coil with respect to a magnetic field generates a voltage	Electro dynamic generator
8	Flow rate	Pressure	Differential pressure is generated between the main pipe line and throat of the Venturi meter/Orifice meter	Venturi meter/orifice meter
9	Flow velocity	Resistance change	Resistance of a thin wire/film is varied by convective cooling in stream of gas/liquid flows	Hot wire anemometer (gas flows) Hot film anemometer (liquid flows)
10	Pressure	Movement of liquid column	The impressed pressure is balanced by the pressure generated by a column of liquid	Manometer
11	Pressure	Displacement	The application of pressure causes displacement in elastic elements	Bourdon gage
12	Gas pressure	Resistance change	Resistance of a heating element varies by convective cooling	Pirani gage

(continued)

Table 1.1 (continued)

S. no.	Input variable to transducer	Output variable of transducer	Principle of operation	Type of device (sensor)
13	Force	Displacement	The application of force against a spring changes its length in proportion to the applied force	Spring balance
14	Force/torque	Resistance change	The resistance of metallic wire or semiconductor element is changed by elongation or compression due to externally applied stress	Resistance strain gage
15	Force	Voltage	An e m f is generated when external force is applied on certain crystalline materials such as quartz	Piezo-electric device
16	Liquid level/thickness	Capacitance change	Variation of the capacitance due to the changes in effective dielectric constant	Dielectric gage
17	Speech/Music/noise	Capacitance change	Sound pressure varies the capacitance between a fixed plate and a movable diaphragm	Condenser microphone
18	Light	Voltage	A voltage is generated in a semiconductor junction when radiant energy stimulates the photoelectric cell	Light meter/solar cell
19	Light radiations	Current	Secondary electron emission due to incident radiations on the photosensitive cathode causes an electronic current	Photomultiplier tube
20	Humidity	Resistance change	Resistance of a conductive strip changes with the moisture content	Resistance hygrometer
21	Blood flow/any other gas or liquid or two phase flow	Frequency shift	The difference in the frequency of the incident and reflected beams of ultrasound known as Doppler's frequency shift is proportional to the flow velocity of the fluid	Doppler frequency shift Ultrasonic flow meter

imprecise, varying as they do from one person to the next. Therefore, there has been a progressive movement toward measurement units that are defined much more accurately.

In the history of measurement, units of measurement occupied an important role. Units are labels which distinguish one type of measurable quantity from other types. Temperatures, mass, time, length, and volume are diverse physical quantities, and therefore have different unit names, degrees, kilograms, seconds, meters, and liters [1].

Many different units of common physical quantities have been used throughout history. Units are necessary so that one person can correctly interpret a measurement another has made, and measuring instruments must be correctly calibrated to give readings in the appropriate units.

Measurement is the process by which one can convert physical parameters into meaningful numbers. The measuring process is one in which the property of an object or system under consideration is compared to an accepted standard unit, a standard defined for that particular property.

1.4.1 Base Quantities and Units

In the SI system of units, there are *seven* base quantities and corresponding units as shown in Table 1.2 [2].

1.4.2 Derived Quantities and Units

These are physical quantities which are resulting from the seven base quantities by mathematical operations such as multiplication, division. Their units are similarly derived as products or quotients of the seven base units as shown in Table 1.3 [3].

Since physical quantities can take a wide range of values, prefixes such as *kilo*, *centi,* and *milli* are used together with units to simplify the expressions for both very large and very small quantities. Table 1.4 lists the decimal multiples and submultiples.

Table 1.2 Unit (base quantity and unit)

Base quantity		SI Base Unit	
Name	Usual symbol	Name	Symbol
Length	l	Meter	M
Mass	m	Kilogram	Kg
Time	t	Second	s
Amount of substance	n	Mole	mol
Electric current	I	Ampere	A
Temperature	T	Kelvin	K
Luminous intensity	l_v	Candela	cd

Table 1.3 Derives quantity and Unit

Quantity	Symbol	Expressed in terms of SI base units	Unit	Unit symbol
Frequency	F	T^{-1}	Hertz	Hz
Velocity	V	Ms^{-1}	Meter per second	m/s
Angular velocity	ω	$m\ m^{-1}\ s^{-1} = s^{-1}$	Radian per second	Rad/s
Acceleration	A	$m\ s^{-2}$	Meter per second square	m/s^2
Angular acceleration	$\propto$	$m\ m^{-1}\ s^{-2} = s^{-2}$	Radian per second square	Rad/s^2
Force	F	$m\ kg\ s^{-2}$	Newton	N ($kg/m/s^2$)
Pressure	p	$m^{-1}\ kg\ s^{-2}$	Newton Per meter Square	N/m^2
Energy	W	$m^2\ kg\ s^{-2}$	Joule	J (Nm)
Charge	Q	s A	Coloumb	S/A
Potential difference	V	$m^2\ kg\ s^{-3}\ A^{-1}$	Volt	V (W/A)
Electric field strength	E	$m\ kg\ s^{-3}\ A^{-1}$	Volt per meter	V/m
Magnetic flux	f	$m^2\ kg\ s^{-2}\ A^{-1}$	Weber	Wb
Magnetic flux density	B	$kg\ s^{-2}\ A^{-1}$	Tesla	(wb/m^2)
Luminance	L	$m^{-2}\ cd$	Lux	lm/m^2
Electric resistance	Ω	$M^2\ kg\ s^{-3}\ A^{-2}$	ohm	V/A
Capacitance	F	$m^{-2}\ kg^{-1}\ s^4\ A^2$	Farad	C/V
Electric conductance	S	$m^{-2}\ kg^{-1}\ s^3\ A^2$	Siemens	A/V
Inductance	H	$m^2\ kg\ s^{-2}\ A^{-2}$	Henry	Wb/A
Phase angle	r	$m\ m^{-1}$	Radian	1

Table 1.4 Decimal multiples and sub-multiples

Prefix	Multiple	Symbol	Example
Atto	10^{-18}	a	1 am (size of electrons)
Femto	10^{-15}	f	Diameter of proton
Pico	10^{-12}	p	10 pF (capacitance)
Nano	10^{-9}	n	1,000 pF (capacitance)
Micro	10^{-6}	μ	1 μF (capacitance)
Mili	10^{-3}	m	mm (100 mm–diameter)
Centi	10^{-2}	c	cm (1 cm distance)
Deci	10^{-1}	dB	Audio level
Deca	10^{1}	dc	Audiologist(decagram)
Hecto	10^{2}	h	Hectoliter (bulk liquid), hectometer (radio band)
Kilo	10^{3}	k	100 kg (weight/mass)
Mega	10^{6}	M	100 MHz (frequency value)
Giga	10^{9}	G	100 GB (hard disk capacity)
tera	10^{12}	T	1 TB (hard disk capacity)

1.5 Classification of Measurement

The classification is important in understanding how different measurement systems work and comparing their performance later on. The benefits of classification are obvious in many walks of life. The animal kingdom, for example, is classified into different species to make it easier to understand and in order to identify common features. Classification also allows us to draw general conclusions and rules about the various classes [4].

A measurement provides a means of describing various phenomena in quantitative terms. It has been quoted "whatever exists, it has to be exists in some amount". The determination of the amount is measurement all about. There are innumerable things in nature which have amounts. The determination of their amounts constitutes the subject of mechanical measurements. The measurements are not necessarily carried out by purely mechanical means. Quantities like pressure, temperature, displacement, fluid flow and associated parameters, acoustics and related parameters, and fundamental quantities like mass, length, and time are typical of those which are within the scope of mechanical measurements. However, in many situations, these quantities are not measured by purely mechanical means, but more often are measured by electrical means by transducing them into an analogous electrical quantity [4].

Classification of measurement allows general rules and conclusions to be drawn about measurement systems in terms of their accuracy, ease of use, and suitability for various types of application.

Methods of measurements:

The methods of measurements may be broadly classified into two categories:

1. Direct method
2. Indirect method

Direct method:

In these methods, the unknown quality (also called the measurand) is directly compared against a standard. The result is expressed as a numerical number and a unit. The standard in fact is a physical embodiment of a unit. In simple words "Measurements are directly obtained". All the physical dimensions are generally measured directly. In the direct measurement, the meaning of measurement and the purpose of the processing operation are identical.

Example: Vernier calipers (as shown in Fig. 1.2), Micrometer etc.

Fig. 1.2 Vernier calipers

Indirect method:
Measurement by direct method is not always possible, feasible, and practicable. Here the meaning of the measurement and the purpose of the processing operation are not identical but are related to each other. This method is used where direct method is not applicable, and the measuring parameter has got some correction with the desired parameter that could not be measured directly. These methods, in most of the cases, are inaccurate because they involve human factors. They are also less sensitive. In simple words "it's obtained by measuring other quantities".

In engineering applications measurement systems are used. These measurement systems use indirect methods for measurement purposes.

1.6 Instrument and Measurement Systems

Measurement involves the use of instruments as a physical means of determining quantities or variables. The instrument serves as an extension of human ability and enables the man to determine the value of unknown quantity or variable which his unaided human abilities cannot measure, a measuring instrument exists to provide information about the physical values of some variable being measured. Because of modular nature of the elements within it, it is common to refer the measuring instrument as a measurement system.

1.7 Classification of Instruments

- **Deflection Type and Null Type:**
  A deflection type instrument is that in which the physical effect generated by the measuring quantity produces an equivalent opposing effect in some part of the instrument which in turn is closely related to some variable like mechanical displacement or deflection in the instrument.
  For example, the unknown weight of an object can be easily acquired by the deflection of a spring sourced by it on the spring balance as shown in Fig. 1.3. Similarly, in a common Bourdon gage, the pressure to be measured acts on the C-type spring of the gage, which deflects and produces an internal spring force to counter balance the force generated by the applied pressure. Deflection instruments are simple in construction and operation. In the deflection type measurement systems, the value of the variable to be measured is shown immediately on a scale or reading.
  An alternative type of pressure gage is the deadweight gage shown in Fig. 1.4, which is a null-type instrument. Here, weights are put on top of the piston until the downward force balances the fluid pressure. Weights are added until the piston reaches a datum level, known as the null point. Pressure measurement is made in terms of the value of the weights needed to reach this null position.

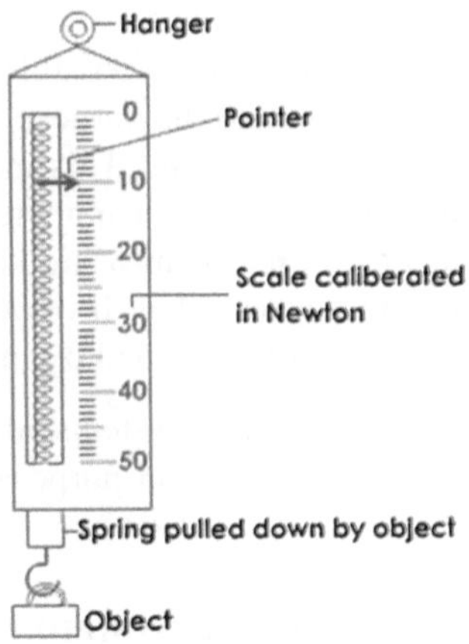

Fig. 1.3 A typical spring balance—a deflection type weight measuring instrument

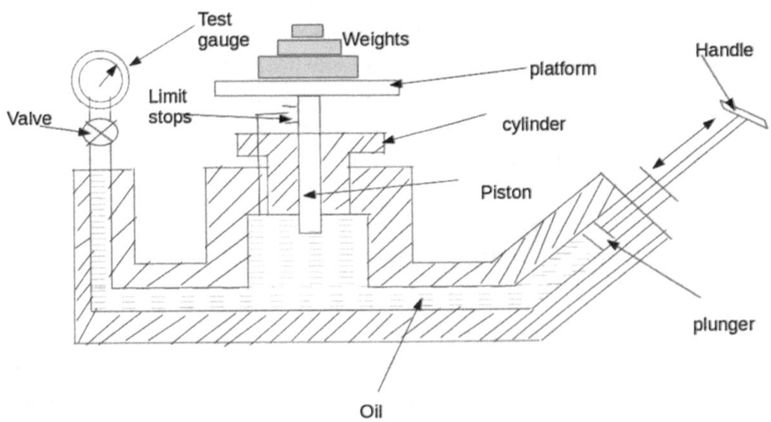

Fig. 1.4 Dead weight pressure gage

The accuracy of these two instruments depends on different things. For the first one it depends on the linearity and calibration of the spring, whilst for the second it relies on the calibration of the weights. As calibration of weights is much easier than careful choice and calibration of a linear characteristic spring, this means that the second type of instrument will normally be the more accurate. This is in accordance with the general rule that null-type instruments are more accurate than deflection types.

In terms of usage, the deflection type instrument is clearly more convenient. It is far simpler to read the position of a pointer against a scale than to add and subtract weights until a null point is reached. A deflection-type instrument is therefore the one that would normally be used in the workplace. However, for calibration duties, the null-type instrument is preferable because of its superior accuracy. The extra effort required to use such an instrument is perfectly acceptable in this case because of the infrequent nature of calibration operations.

- **Manually Operated Type and Automatic Type:**
  Any instrument which requires the services of human operator is a manual type of instrument. The instrument becomes automatic if the manual operation is replaced by an auxiliary device incorporated in the instrument. An automatic instrument is usually preferred, because the dynamic response of such an instrument is fast and also its operational cost is considerably lower than that of the corresponding manually operated instrument.
  Manually operated measurement systems require the intervention—a form of human intervention for their operation. This can be costly, and will affect the speed of response of the system. If such systems can be automated, then human intervention can be dispensed with, resulting in cost reduction and enhancement of speed. Most null-type devices are manually operated, and most deflection devices are automatic.
- **Analog Type and Digital Type:**
  Analog instruments are those that present the physical variables of interest in the form of continuous or step-less variations with respect to time. These instruments usually consist of simple functional elements. Therefore, the majority of present-day instruments are of analog type as they generally cost less and are easy to maintain and repair.
  An analog instrument gives an output that varies continuously as the quantity being measured changes. The output can have an infinite number of values within the range that the instrument is designed to measure.
  On the other hand, digital instruments are those in which the physical variables are represented by digital quantities which are discrete and vary in steps. Furthermore, each digital number is a fixed sum of equal steps which is defined by that number. The relationship of the digital outputs with respect to time gives the information about the magnitude and the nature of the input data.
- **Indicating Instruments and Instruments with a Signal Output:**
  The measurement system could just provide an output indication, or the output could be fed it into a control system (e.g., for negative feedback). The signal output could be a voltage or current or an optical signal or a pneumatic signal. The problem with indicating type measurement system is the need for human intervention. Humans make mistakes when reading scales.
  The class of indicating instruments normally includes all null-type instruments and most passive ones. Indicators can also be further divided into those that have an analog output and those that have a digital display. A common analog indicator is the liquid-in-glass thermometer.
  Instruments that have a signal-type output are commonly used as part of automatic control systems. In other circumstances, they can also be found in measurement systems where the output measurement signal is recorded in some way for later use. Usually, the measurement signal involved is an electrical voltage, but it can take other forms in some systems such as an electrical current, an optical signal, or a pneumatic signal.

- **Active Type and Passive Type:**
  Instrument systems are divided into active or passive ones according to whether the instrument system output is entirely produced by the quantity being measured or the quantity being measured simply modulates the magnitude of some external power source.
  In active instruments system, the external power source is usually in electrical form, but in some cases, it can be other forms of energy such as a pneumatic air supply, electronic supply, or hydraulic supply for their operation.
  In self-generating (or passive) instruments, the energy requirements of the instruments are met entirely from the input signal.
- **Contacting Type and Non-Contacting Type:**
  A contacting type of instrument is one that is kept in the measuring medium itself. A clinical thermometer is an example of such instruments. On the other hand, there are instruments of noncontacting or proximity type. These instruments measure the desired input, even though they are not in close contact with the measuring medium.
  For example, an optical pyrometer monitors the temperature of, say, a blast furnace, but is kept out of contact with the blast furnace. Similarly, a variable reluctance tachometer, which measures the rpm of a rotating body, is also a proximity type of instrument.
- **Smart and Non-Smart (dumb) Measurement Systems:**
  Measurement systems having on-board intelligence (usually containing microprocessors) are called smart measurement systems and are capable of storing data, processing it, and sending alarms. It can have its own serial bus to connect to other microprocessors.
  A dumb/nonsmart or conventional instrument is that in which the input variable is measured and displayed, but the data are processed by the observer. For example, a Bourdon pressure gage is termed as a dumb instrument because though it can measure and display a car tyre pressure, but the observer has to judge whether the car tyre air inflation pressure is sufficient or not.
  An intelligent or smart instrument may include several or all like the output of the transducer in electrical form, the output of the transducer should be in digital form, interface with the digital computer, elements for noise reduction, error estimation, self-calibration, gain adjustment, etc., and elements for the output driver for suitable digital display.

1.8 Transducer

A transducer is a mechanism that transfers one form of energy or physical quantity into another, in accordance with some defined relationship. Where a transducer is the sensing element, which responds directly to the physical quantity to be measured and this forms part of an instrumentation or control system, then the transducer is often referred to as a sensor.

Transducer is an electronic device that is proficient to convert one form of energy into another. Many devices around us from the light bulb to the cell phone to a computer screen are transducers.

A transducer mainly lists in one of the following categories:

- Electromagnetic
- Thermoelectric
- Electrochemical
- Electro acoustic
- Electromechanical
- Photoelectric

1.8.1 Transducer Characteristics

Transducers are classified according to the physical quantity they measure (e.g., temperature, force etc.). Beyond the obvious selection of the type of transducer required to measure a particular physical quantity and any cost considerations, the characteristics which are most important in determining a transducers applicability for a given application are as follows:

- Accuracy
- Sensitivity
- Repeatability
- Range
- Precision

1.8.1.1 Accuracy

Accuracy indicates the closeness of the measured value with the actual or true value, and is expressed in the form of the *maximum error (= measured value—true value)* as a percentage of full scale reading. It is always better to choose a scale of measurement where the input is near full-scale value. But the true value is always difficult to get. In other words, it is a degree of conformity of the measured value with the standard or ideal value. Also it is expressed as percentage of the sensor range. *In simple words, it represents how close a reading is to the true value.*

1.8.1.2 Sensitivity

It can be defined as the ratio of the *incremental output* and the *incremental input*. As defining the sensitivity, suppose that the input–output characteristic of the instrument is approximately linear in that range. Sensitivity of a spring balance can be expressed as 25 mm/kg (say), indicating additional load of 1 kg will cause additional displacement of the spring by 25 mm.

Again sensitivity of an instrument may also vary with temperature or other external factors. This is known as *sensitivity drift*. Suppose the sensitivity of the spring balance mentioned above is 25 mm/kg at 20 °C and 27 mm/kg at 30 °C. Then the sensitivity drift/ °C is 0.2 (mm/kg)/ °C. In order to avoid this type of sensitivity drift, instruments are either kept at controlled temperature, or appropriate integral temperature compensation scheme provided inside the instrument.

1.8.1.3 Repeatability

It is defined as closeness of conformity between a numbers of repeated measurements of the same variable (value) under the same operating conditions, which is approaching in the same direction.

1.8.1.4 Range (or Span)

It defines the maximum and minimum values of the inputs or the outputs for which the instrument is recommended to be used.

Most sensors have limited range over which a process variable can be measured, defined by the lower and upper range values. Generally, larger the range is in result of the poorer accuracy and reproducibility.

Example: A temperature measuring Instrument input range may be 100–500 °C and the output range may be 4–20 mA. Most sensors have limited range over which a process variable measured is define by lower and higher range value.

1.8.1.5 Precision

Precision indicates the repeatability or reproducibility of an instrument (but does not indicate accuracy). If an instrument is used to measure the same input, but at different instants, spread over the whole day, successive measurements may vary randomly.

The random fluctuations of readings, is often due to random variations of several other factors which have not been taken into account, while measuring the variable. A precise instrument indicates that the successive reading would be very close, or in other words, the standard deviation σ_e of the set of measurements would be very small. In Simple words, it indicates how close a reading is to the Average value.

1.9 Measurement Techniques

Measurement involves the use of instruments as a physical means of determining quantities or variables. Number of instruments are available to be used for measuring the same physical quantity as shown in Table 1.5 and schematic of measurement systems is shown in Fig. 1.5.

Number of techniques are available to measure the following physical quantities:

- Flow
- Pressure
- Temperature
- Level
- Conductivity
- pH
- Density
- Viscosity
- Voltage
- Radiation
- Inductance
- Capacitance
- Resistivity
- Frequency
- Chemical composition
- Various physical properties

Table 1.5 Physical quantity measurement

Physical quantity	Instrument
Length	Ruler, vernier calipers, and micrometer
Mass	Electronic balance, lever balance
Weight	Spring balance, Newton meter
Time	Stopwatch, calibrated time base of a CRO
Temperature	Resistance thermometer, thermocouple, and RTD
Electric current	Galvanometer, ammeter
Potential difference	Voltmeter, digital multimeter

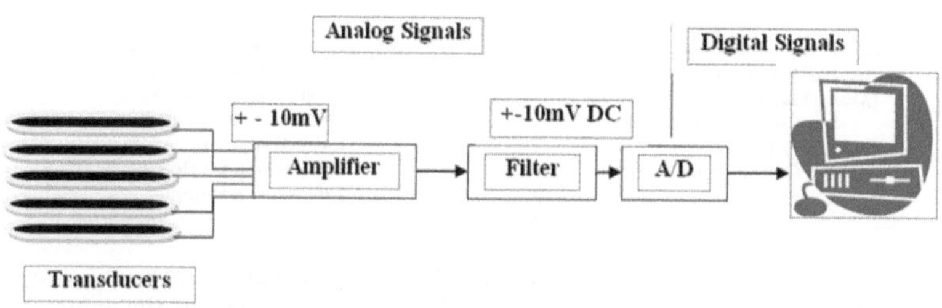

Fig. 1.5 Measurement systems

1.9.1 Flow Measurement

Accurate measurement of flow rate of liquids and gases is an essential requirement for maintaining the quality of industrial processes. In fact, most of the industrial control loops control the flow rates of incoming liquids or gases in order to achieve the control objective. As a result, accurate measurement of flow rate is very important. Various techniques are available for measuring flow. Each technique has its advantages and disadvantages, and selecting a proper technique depends on its specific applications [5].

Needless to say that there could be diverse requirements of flow measurement, depending upon the situation. It could be volumetric or mass flow rate, the medium could be gas or liquid, the measurement could be intrusive or nonintrusive, and so on. As a result, different types of flow measuring techniques are used in industries. The common types of flow meters that find industrial applications can be listed as below in Fig. 1.6.

1.9.2 Pressure Measurement

Measurement of pressure inside a pipeline or a container in an industrial environment is a challenging task, keeping in mind that pressure may be very high, or very low (vacuum); the medium may be liquid, or gaseous. Pressure is a force generated inside a sealed volume by the atoms and molecules either pressing out (positive pressure) or pulling in (negative pressure).

Different methods are available for measurement of pressure and force. Elastic elements, namely diaphragms and bourdon tubes are mainly used for pressure measurement. On the other hand, strain gages are commonly used for measurement of force. These elastic elements change their shape with applied pressure and the change of shape can be measured using suitable deflection transducers.

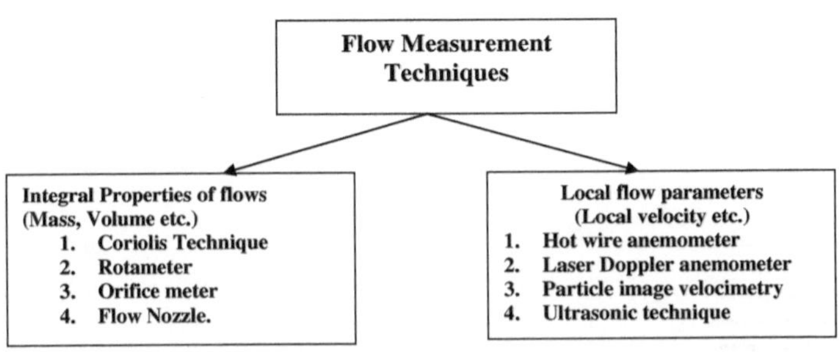

Fig. 1.6 Types of flow meters. **a** Obstruction type (differential pressure or variable area). **b** Inferential (turbine type). **c** Electromagnetic. **d** Positive displacement (integrating). **e** Fluid dynamic (vortex shedding). **f** Anemometer. **g** Ultrasonic. **h** Mass flow meter (Coriolis)

Pressure is defined as force acting evenly over a unit area.

$$\text{Pressure} = \text{Force}/\text{Unit Area} = F/A$$

This force can be exerted by liquids, by gases or vapors, or by solid bodies. Surface compression takes place at the interface between two solid bodies, but for our purposes we can consider this additional force negligible.

Pressure can also be uttered in terms of metric (SI) units. The basic metric unit of force is the Newton (N) and the basic unit of pressure is the Pascal (Pa).

A pressure measurement can be explained as either static or dynamic. In conducting tests where no motion is taking place is referred to as *static* pressure. Examples of static pressure include the pressure of the air inside a balloon or water inside a basin. Often times, the motion of a fluid changes the force applied to its surroundings. Such a pressure measurement is known as dynamic pressure measurement. For example, the pressure inside a balloon or at the bottom of a water basin would change as air is let out of the balloon or as water is poured out of the basin.

There are three methods for measuring pressure: absolute, gage, and differential. *Absolute pressure* is referenced to the pressure in a vacuum, whereas *gage and differential pressure*s are referenced to another pressure such as the ambient atmospheric pressure or pressure in an adjacent vessel. Pressure is measured by converting the physical phenomenon to an intermediate form, such as displacement, which can be measured by a transducer.

Diaphragms, Bourdon tube, Manometers, Bellows, Capacitive pressure sensor, Fiber-optic pressure sensors, Resonant-wire devices, Dead-weight gage, Thermocouple gage, Pirani gage, McLeod gage, and Ionization gage are types of sensors/devices available for measuring pressure.

Pressure measurement is vital and essential in following industrial applications:

- Drilling technology utilizes pressure sensors for real-time down hole data transfer
- Weather forecasting
- Medicine
- Aviation
- Pressure Vessels

1.9.3 Temperature Measurement

The need for reliable temperature scales as a prerequisite for clear measurement results was already recognized in the early stages of thermometer development. Therefore, efforts were made to calibrate the thermometers on reference points with good reproducibility. Since it was easy to determine the freezing point and boiling point of water at the time, these reference points were quickly used for these purposes. In 1694, *Renaldini* recommended the boiling point of water as the upper limit of the scale.

The word *temperature* was coined to describe the degree of hotness or coolness of a material or a body. The thermometer—a temperature measuring instrument was first developed.

Temperature is a quantity that can be readily perceived by the human senses, by contact with the skin or by visually by eye. The concepts "warm", "cold", "ice cold" or at high temperatures "red hot" or "white hot" are known to everyone. Human beings are relatively good at realizing temperature differences, whether something is colder or warmer than another substance or object.

From the physical point of view, temperature can be described as a measure of the energy inherent in a body, which results from the unordered movement of its atoms or molecules. Temperature is a state variable, which together with quantities such as mass, heat capacity and others, describes the energy content of a body or system. Therefore, temperature could be measured directly in energy units. However, the tradition of specifying the temperature in degrees had already been introduced far earlier and was well established in physics, so that for practical reasons it was not reasonable to discontinue the use of degrees.

Temperature plays an important role in the reproducibility of product quality, the profitability of production processes and the operational reliability of industrial installations and equipment. Temperature measurement methods can basically be subdivided into contact and noncontact temperature measuring methods. In contact temperature measuring methods, the thermometer, and in particular the part containing the temperature—sensitive sensor, is brought into thermal contact with the measured medium at the area of installation. The energy exchange between thermometer and measured medium takes place primarily via thermal conduction.

Noncontact temperature measuring methods use the energy exchange between measured object and thermometer by heat radiation.

Temperature sensors based on different principles have been developed. They can be broadly classified into the following groups:

1. Liquid and gas thermometer
2. Bimetallic strip
3. Resistance thermometers (RTD and Thermistors)
4. Thermocouple
5. Junction semiconductor sensor
6. Radiation pyrometer

1.9.4 Displacement and Speed Measurement

Displacement and speed are two important parameters whose measurements are important in many position and speed control schemes. Error free measurements of these two parameters are required in order to accomplish good control performance [5, 6].

The physician and biomedical researcher are interested in measuring the size, shape, and position of the organs and tissues of the body. Variations in these parameters are important in discriminating normal from abnormal function. Displacement sensors can be used in both direct and indirect systems of measurement. Direct measurements of displacement are used to determine the change in diameter of blood vessels and the changes in volume and shape of cardiac chambers.

Indirect measurements of displacement are used to quantify movements of liquids through heart valves. An example is the movement of a microphone diaphragm that detects the movement of the heart indirectly and the resulting heart murmurs.

Displacement measurement can be of two types: contact and noncontact types. Besides the measurement principles can be classified into two categories: electrical sensing and optical sensing. In electrical sensing, passive electrical sensors are used and variation of either inductance or capacitance with displacement is measured. On the other hand, the optical method mainly works on the principle of intensity variation of light with distance.

Potentiometer, LVDT, piezo electric transducer, inductive and capacitive type measurement, tachogenerator, gagnetic pickup, and proximity are mostly used sensors for displacement and speed measurement.

1.9.5 Level Measurement

A level measurement is an integral part of process control, and may be used in a variety of industries [6].

- Level measurement may be divided into two categories:
  - Point level measurement
  - Continuous level measurement
- Point level sensors are used to mark a single discrete liquid height, a preset level condition (as a high alarm or a low alarm condition).
- Continuous level sensors provide an analog output that directly correlates to the level within the containing vessel.

This analog signal may be directly linked to a visual indicator or to a process control loop, forming a level management system.

There are several instances where we need to monitor the liquid level in vessels. In some cases the problem is simple, we need to monitor the water level of a tank; a simple float type mechanism will suffice. But in some cases, the vessel may be sealed and the liquid a combustible one; as a result, the monitoring process becomes more complex. Depending upon the complexity of the situation, there are different methods for measuring the liquid level, as can be summarized as follows:

1. Float type
2. Hydrostatic differential pressure gage type
3. Capacitance type
4. Ultrasonic type
5. Radiation technique

1.9.6 Humidity Measurement

Humidity measurement finds wide applications in different process industries. Moisture in the atmosphere must be controlled below a certain level in many manufacturing processes, e.g., semiconductor devices, optical fibers etc. Humidity inside an incubator must be controlled at a very precision level. Textiles, papers, and cereals must be dried to a standard storage condition in order to prevent the quality deterioration. Museums, clean rooms, paint booths, operating rooms, hospitals, and test chambers all must monitor and document humidity and temperature conditions.

Humidity is defined as the water vapor content in air or other gases. In normal air there is typically about 1 % of water vapor, but it is generally present in greater or lesser amounts. High humidity makes hot days feel even hotter. Low humidity can give people a feeling of a dry throat, or sensations of "static" when touching things. Humidity is measured using a hygrometer.

Humidity is usually measured in terms of absolute humidity (the ratio of the mass of water vapor to the volume of air or gas), dew point (the temperature and pressure at which a gas begins to condense into a liquid), and relative humidity, or RH (the ratio of the moisture content of air compared to the saturated moisture level at the same temperature or pressure).

The output of all absorption-based humidity sensors (capacitive, resistive, conductive film, etc.) is affected by both temperature and %RH. Because of this, temperature compensation is used in applications that term for either higher accuracy or wider operating temperature ranges.

Humidity can be measured in different ways. Popular are Hygrometer and Psychrometer. The most common humidity semiconductor type sensors are capacitive, resistive, and thermal conductivity.

1.9.7 Measurement of pH

pH is a measure of hydrogen ion concentration in aqueous solution. It is an important parameter to determine the quality of water. pH is a unit of measure which expresses the degree of acidity or alkalinity of a solution. It is measured on a scale of 0–14 as shown in the Fig. 1.7. The term pH is derived from "p", the mathematical symbol of the negative logarithm, and 'H' the chemical symbol of Hydrogen. The formal definition of pH is the negative logarithm of the hydrogen ion activity.

$$\mathrm{pH} = -\log[\mathrm{H}^+]$$

Pure water dissociates to yield 10^{-7} M of $[H^+]$ and $[OH^-]$ at 25 °C; thus, the pH of pure water is neutral, i.e., 7.

$$\mathrm{pH}_{\mathrm{water}} = -\log[\mathrm{H}^+] = -\log 10^{-7} = 7$$

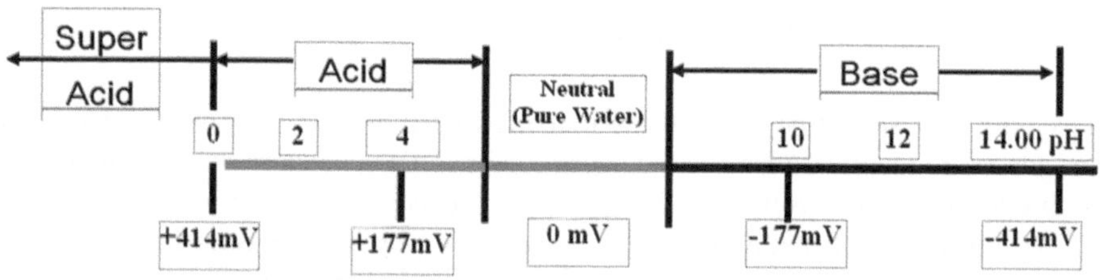

Fig. 1.7 pH scale

Solutions with a higher [H⁺] ions than water (pH less than 7) are acidic; solutions with a lower [H⁺] than water (pH greater than 7) are basic or alkaline.

Measuring pH evaluates the potential of solutions with unknown [H⁺] to a known reference potential. PH meters alter the voltage ratio between a reference half-cell and a sensing half-cell to pH values. In acidic or alkaline solutions, the voltages on the outer membrane surface vary proportionally to changes in [H⁺]. The pH meter classifies the change in potential and determines [H⁺] of the unidentified by the Nernst equation:

$$E = E_o + (2.3\,RT)/nF\log\,\{\text{unknown}\,[H^+]/\text{internal}\,[H^+]\}$$

where E = full potential difference (measured in mV); E_o = reference potential; R = gas constant; T = temperature in Kelvin; n = number of electrons; F = Faraday's constant; $[H^+]$ = hydrogen ion concentration.

pH measurement is measured by using pH meter. A pH measurement system consists of three parts:

1. Basic electrode
2. Reference electrode
3. Impedance measurement meter

The actual pH of the solution changes with the temperature change.

1.9.8 Rotational Motion Measurement

The various devices available for measuring rotational displacements are: Rotational displacement transducers measure the angular motion of a body about some rotation axis. Circular and helical potentiometers, rotational differential transformer, incremental shaft encoders, coded-disk shaft encoders, the resolver, the synchro, the induction potentiometer, and the rotary inductosyn, Gyroscopes.

1.9.9 Resistance Measurement

It is normally seen that methods which involve the measurement of change in resistance are ideal to those use other principles. This is because alternating as well as direct current and voltages are suitable for resistance measurement [7].

The resistance of metal conductor is expressed by an equation that takes up a few physical quantities. The relationship is $R = \rho L/A$; where

R Resistance, Ω
L Length of conductor, m
A Cross-sectional area of conductor, m^2 and
ρ Resistivity of conductor material, Ω-m

Any method of changeable one of the extent involved in the relationship can be the design basis of resistance measurement sensor. The standard devices and methods available for measuring change in resistance, which is measured in units of *ohms*, include the DC bridge circuit, the voltmeter–ammeter method, the resistance–substitution method, the digital voltmeter and the ohmmeter. Devices that convert the measured quantity into a change in resistance include the resistance thermometer, thermistor, the wire-coil pressure gage, and the strain gage.

1.9.10 Power Measurement

Power may be defined as the rate at which energy is transformed or made available. The power in a circuit at any instant is equal to the product of the current in the circuit and the voltage across its terminals at that instant [8, 9].

In a DC circuit if V is voltage and I is the current then power is given by,

$$P = V \cdot I$$

If the circuit resistance is known power may be calculated from one of the equivalent formula,

$$P = V^2/R = I^2R$$

In an AC circuit, the instantaneous power varies continuous as the current and voltage go through a cycle of values. Power in an AC circuit is given by,

$$P = V \cdot I \cdot \cos\phi$$

V RMS Value of AC voltage
I RMS value of AC current
ϕ phase angel between voltage and current, and
$\cos\phi$ power factor to the AC current

Ammeter and voltmeters, induction wattmeter, electrodynamometer, and AF power output meter are used for power measurement.

1.9.11 Inductance Measurement

Inductance is the characteristic of an electrical circuit that opposes the initial, prevent, or a change in value of current [8, 9]. The main device that has an output in the form of a change in inductance is the inductive displacement sensor. Inductance is measured in *henry* (H). *Inductance is the link between electric circuits and magnetic fields.*

An inductor is a coiled conductor. It is a device to accumulate energy in a magnetic field. An inductor consists of a wire wound around a core material. Air is the simplest inductor core material because it is constant, however for physical efficiency, magnetic materials such as iron and ferrites are commonly used. The inductor's core material, its length, and number of turns directly affect the inductor's ability to carry current.

Inductor measurements can be made in series or parallel mode. In the case of a large inductance, reactance at a given frequency is relatively large, so parallel reactance becomes more significant than series reactance. A parallel equivalent circuit should be employed to evaluate a large inductance.

Inductance is a basic electrical property of any coil. The inductance of a coil depends on the number of turns, diameter of the coils, the length of the coil, and the nature of the core. To get an accurate inductance measurement, the inductor must be tested under actual conditions with current flowing through the coil.

1.9.12 Capacitance Measurement

A capacitor is basically consists of two conductors separated by a dielectric medium. The variable to be measured will cause an effect either by increasing the distance between two plates or by changing the dielectric constant [8, 9]. Capacitance of parallel plate capacitor (Fig. 1.8) whose plates are displaced by a distance d is given as:

$$C = \varepsilon_0 \varepsilon_r A/d$$

where A is the area of cross-section of the plates, $\varepsilon_0 \varepsilon_r$ are absolute and relative dielectric constants of the medium.

Any change in A and d will change the capacitance C. However, a change in separation d between the plates is simple and is widely used to sense the displacement.

Devices having an output in the form of a change in capacitance include the capacitive level gauge, the capacitive displacement sensor, the capacitive

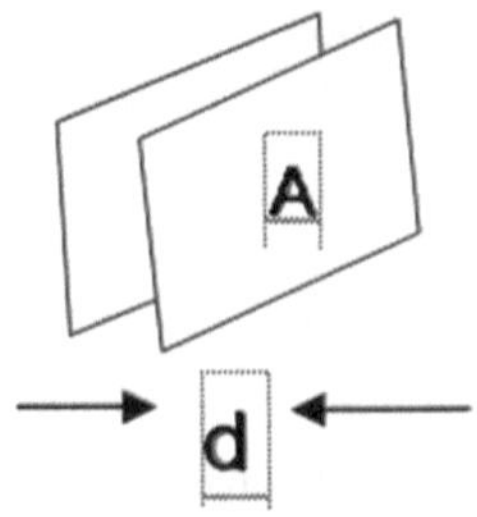

Fig. 1.8 Parallel Plate

moisture meter, and the capacitive hygrometer. Capacitance is measured in units of *Farads* (F).

Capacitance measurement is suitable for measuring the level of nonconductive, i.e., dielectric liquids like oils, gasoline, or liquid gases for corrosive acids and in high pressure processes.

Capacitive sensors have a wide variety of uses. Some are:

- Flow–Many types of flow meters convert flow to pressure or displacement, using an orifice for volume flow or Coriolis Effect force for mass flow. Capacitive sensors can then measure the displacement.
- Pressure–A diaphragm with stable deflection properties can measure pressure with a spacing–sensitive detector.
- Liquid level–capacitive liquid level detectors sense the liquid level in a reservoir by measuring changes in capacitance between conducting plates which are immersed in the liquid, or applied to the outside of a nonconducting tank.
- Spacing–If a metal object is near a capacitor electrode, the mutual capacitance is a very sensitive measure of spacing.
- A capacitive type sensor is placed on bed to record respiratory movements from the lungs and ballistrographic movements from the heart. Capacitors formed by placing mica insulators between corrugated metal layers can be used to measure pressure between the foot and shoe.

1.9.13 Frequency Measurement

An important electrical quantity with no equivalent in DC circuits is *frequency*. Frequency measurement is very essential in many applications of alternating current, particularly in AC power systems designed to run efficiently at one frequency only. If an electromechanical alternator is generating the AC, the frequency will be directly proportional to the shaft speed of the machine, and frequency could be measured simply by measuring the speed of the shaft. If frequency needs to be measured at some distance from the alternator, though other means of measurement will be necessary.

Frequency measurement is required as part of those devices that convert the measured physical quantity into a frequency change, such as the variable-reluctance velocity transducer, stroboscopes, the vibrating-wire force sensor, the resonant-wire pressure sensor, the turbine flow meter, the Doppler-shift ultrasonic flow meter, the transit-time ultrasonic flow meter, the vibrating level sensor, the quartz moisture meter, and the quartz thermometer. In addition, the output relationship in some forms of AC bridge circuit used for measuring inductance and capacitance requires accurate measurement of the bridge excitation frequency.

Frequency is calculated in units of *hertz* (Hz). The digital counter timer is the most common instrument for measuring frequency. The oscilloscope is also commonly used for obtaining approximate measurements of frequency.

Appendix A: Definitions of the SI Base Units

kilogram	The mass of a piece of platinum–iridium alloy held in reserve under standard conditions near Paris
second	The duration of 9192613770 periods of radiation equivalent to the transition between the two hyperfine levels of the ground state of the caesiurn-133 atom
metre	Distance traveled in 1/299792458 of a second by plane EM waves in a vacuum
Ampere	The electric current which, if maintained in two straight parallel conductors of infinite length and negligible circular cross-section, when placed one meter apart in a vacuum would produce, per meter of length, a force of 2×10^7 N between the two conductors
Kelvin	The fraction 1/273.16 of the thermodynamic temperature of the triple point of water.
mole	A mole the amount of substance of a system which contains as many molecules, atoms, or elementary entities as there are carbon atoms in 0.012 kg of carbon-12

References

1. Yang Huang, Yani Xu, Yijia Guan, et. al. Quelle : RUDAR - Roskilde University Digital Archive, http://rudar.ruc.dk/bitstream/1800/2496/1/sensor.pdf
2. The International System of Units (SI): Bureau international des Poids et Measures (2010), http://www.bipm.org/en/si/, 30 Nov 2010
3. The International System of Units from NIST: National Institute of Standards and Technology, http://physics.nist.gov/cuu/Units, 30 Nov 2010
4. B.C. Nakra, K.K. Chaudhry, *Instrumentation Measurement and Analysis* (Tata McGraw Hill, NY, 1985)
5. E.O. Doebelin, *Measurement Systems: Application and Design*, 5th edn. (McGraw Hill, NY, 2003)
6. A.S. Morris, *Measurement & Instrumentation Principles* (Elsevier, London, 2001)
7. L.F. Adams, *Engineering Measurements and Instrumentation* (Hodder & Stoughton Ltd, London, 1975)
8. J.J. Carr, *Elements of Electronics Instrumentation and Measurement*, 3rd edn. (Prentice Hall, NJ, 1996)
9. B.A. Gregory, *An Introduction to Electrical Instrumentation and Measurement Systems*, 2nd edn. (McMillan Press, NY, 1981). (ELBS Edition)

Chapter 2
Sensor

2.1 Introduction

The most important element in a measurement system is the sensor. If the data is unclear or stained by the sensor, there is often little that can be done to correct it. Sensors are most commonly used to make scientific measurements, as opposed to qualitative detection or presence sensing. Sensors are available to measure nearly anything you can think of, and many things you would never think of.

There are number of applications where sensors can be utilized, and there are number of sensors to choose from. Choosing the right sensor can be confusing and takes careful thought and planning. Often, more than one sensor will do the work. As the application becomes more complex the more difficult it is to choose the right sensor for a given application. The study of sensors is multidisciplinary and requires a comprehensive knowledge of general physics, solid-state physics, electronics, technology and semiconductor-manufacturing techniques.

A basic instrument system consists of three elements:

1. Sensor or input device
2. Signal processor
3. Receiver or output device

2.2 Sensor

A *sensor* is a device that converts a physical phenomenon (for example: temperature, blood pressure, humidity, speed, etc.) into an electrical signal. As such, sensors represent part of the interfacing between the physical world and the world of electrical devices, i.e., mobile phone. The other part of this interface is represented by *actuators*, which convert electrical signals into physical phenomena. Energy *information* conversion is the objective of a sensor. The information available in one energy form must be converted into the same or another energy

H. K. Patel, *The Electronic Nose: Artificial Olfaction Technology*,
Biological and Medical Physics, Biomedical Engineering,
DOI: 10.1007/978-81-322-1548-6_2,

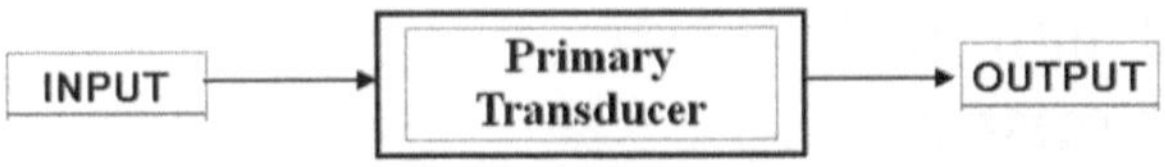

Fig. 2.1 Block diagram of sensor

form, which is exactly the same information content as the originating energy form. Block diagram of sensor is shown in Fig. 2.1.

The sensor or the sensing element is the first element in a measuring system and takes information about the variable being measured and transforms it into a more suitable form to be measured. Figure 2.2 illustrates the difference between sensor and transducer.

Sensor is sometimes called a primary measuring element, it can be found simply as a mercury thermometer to measure the temperature. It may be embedded in the transducer to perform its function. That means the transducer consists of a primary element (sensor) plus a secondary element (signal conditioning circuit) that transforms the passive change or small voltage signal into active signal range that can be easily used in other chains of the control loop.

$$\text{Transducer} = \text{Sensor} + \text{Signal conditioning circuit}$$

The differences between sensor and transducers are very negligible. A sensor performs a transducing action, and the transducer must necessarily sense some physical quantity. The difference lies in the efficiency of energy conversion. The purpose of a sensor is to detect and measure, and its efficiency is 5 or 0.1 % which is almost immaterial, provided the figure is known. A transducer, by contrast is intended to convert energy, and its efficiency is important, though in some cases it may not be high. Linearity of response, defined by plotting the output against the input, is likely to be important for a sensor, but it is of much less significance for a transducer. Example: Fig. 2.3, Transducer configuration (displacement measurement using ultrasonic sensor).

The transducer uses sensor as one of its potential operation elements, whereas a sensor has no internal separate operating elements, and just the sensing element itself. A transducer relies on the sensor to produce the quantity or energy conversion. It is also capable of modifying the sensors output signal.

Sensors and their related interfacing circuits are used to measure various types of physical properties such as temperature, force, pressure, flow, position, light

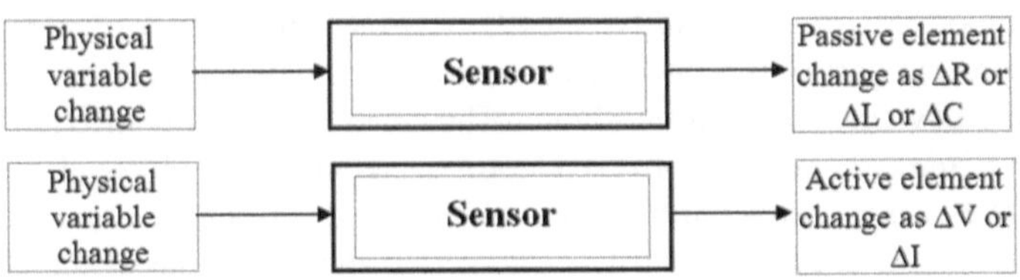

Fig. 2.2 Block diagram of sensor and transducer

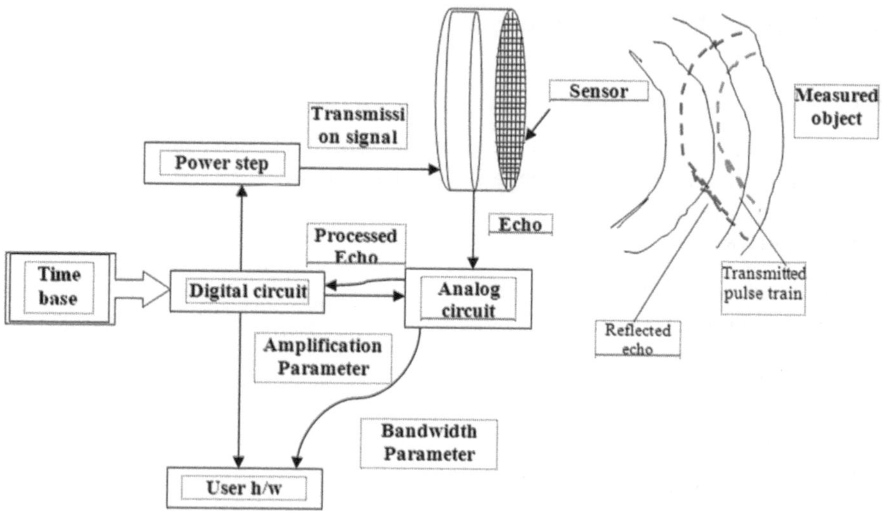

Fig. 2.3 Transducer configuration with the use of sensor

intensity, etc., and the sensor output is conditioned and analyzed to provide the equivalent measurement of the physical property.

2.2.1 Criteria to Choose a Sensor

There are certain features to be considered while choosing a sensor. They are:

1. Accuracy
2. Environmental condition—usually has limits for temperature/humidity
3. Range—Measurement limit of sensor
4. Calibration—Essential for most of the measuring devices as the readings changes with time
5. Resolution—Smallest increment detected by the sensor
6. Cost
7. Repeatability—The reading that varies is repeatedly measured under the same environment

Sensors do not operate by themselves. They are generally part of a larger system consisting of signal conditioners and various analog or digital signal processing circuits. The *system* could be a measurement system, data acquisition system, or process control system. We will not cover all possible types of sensors but discuss popular types of sensors such as: Temperature, Flow, Pressure, Level, displacement, positional, etc.

2.3 What to Look for in a Sensor?

- Dynamic range
  - Minimum and maximum range of the measured physical quantity
  - Minimum and maximum range for electric output
- Input/output relation
- Sensitivity/Resolution
  - Smallest change to be detected
- Power requirements
  - Passive/active
  - Power consumption/requirements
- Band width and frequency response
- Calibration process

2.3.1 Measurand Categories

1. Mechanical quantities related to solid bodies
   - Position
   - Navigation
   - Speed and RPM
   - Acceleration/vibration
   - Tilt
   - Distance
   - Caliper dimension
   - Roughness
   - Thickness
   - Force
   - Torque
2. Mechanical quantities related to fluids (gases, liquids, and powders)
   - Pressure
   - Differential pressure
   - Flow
   - Level
3. Thermal quantities
   - Temperature
   - Infrared signal

4. Optical quantities
   - Light intensity
   - Optical images
   - Other radiation
5. Acoustic and vibration quantities
   - Sound
   - Ultrasonic images
6. Electric and magnetic quantities
   - Electric current
   - Magnetic field
7. Qualities of materials, liquids, and the environment
   - Material quality
   - Liquid quality
   - Environment quality
8. Chemical quantities
   - Humidity sensor
   - Ions in liquid
   - Gases in liquid
   - Chemicals in gas
9. Biological quantities
   - Glucose sensor
   - DNA chips
   - Other biosensor

2.4 Classification of Sensors

- Classification by broad area of detection [1]:
  - Electric
  - Magnetic
  - Electromagnetic
  - Acoustic
  - Chemical
  - Optical
  - Heat, Temperature
  - Mechanical
  - Radiation
  - Biological, etc.

- Classification based on Application:
  - Industrial process control, measurement, and automation.
  - Non-industrial use—Aircraft, Medical products, Automobiles, Consumer electronics, and other types of sensors.
- Classification based on power or energy supply requirement:
  - Active Sensor—Sensors that require power supply are called as Active Sensors. Example: photoconductive cell, Thermistor, Strain Guage, Capacitive and inductive sensor, etc.
  - Passive Sensor—Sensors that do not require power supply are called as Passive Sensors. Example: Radiometers, Thermocouple, Piezo electronic, etc.
- Classification based on Measurand contact:
  - Contact sensor: a sensor that requires physical contact with the stimulus. Examples: strain gauges, most temperature sensors.
  - Non-contact sensor: requires no physical contact. Examples: most optical and magnetic sensors, infrared thermometers, etc.
- Classification based on specifications:
  - Accuracy
  - Sensitivity
  - Stability
  - Response time
  - Hysteresis
  - Frequency response
  - Input (stimulus) range
  - Resolution
  - Linearity
  - Hardness (to environmental conditions, etc.)
  - Cost
  - Size, weight
  - Construction materials
  - Operating temperature, etc.

Elemental transduction processes (ETPs) mechanisms fall into four quite distinct types: energy conversion (C), energy dispersion (D), energy modulation (M), and property modulation (P). Classification of sensory technology by measurand category is shown in Table 2.1 [2].

2.4.1 Temperature Sensor

The word *temperature* was coined to describe the degree of hotness or coolness of a material body. All temperature sensors have nonlinear transfer functions, exceptional is IC sensors. Sensor outputs may be well digitized directly by high

Table 2.1 Sensor technology classification by measurand category [2]

Measurand/input variable	Sensor/transducer technology	Sensing subsystem mechanism	Applications
Mechanical quantities			
Linear position	Encoders (optical)	M-M. Optical probe—pin photodiode	Industrial robots, machine tools
Angular position	Resolvers (magnetic)	M-C. AC magnetic probe-coil	Manufacturing, process control
Position	Potentiometer (variable voltage divider)	C-M. Work done on wiper—voltage probe	Automobile gas pedal, throttle valve
	LVDT—linear variable differential transformer	M-C. AC magnetic probe—coil	Automotive, aerospace, manufacturing
	Variable reluctance	M-C. AC magnetic probe—coil	Angular position
Direction (navigation)	Mechanical Gyros	M-M. Angular momentum—capacitance	Inertial reference systems
Speed and RPM	Tachometer	C. Mechanical to electrical (magnetic induction)	Frequency and amplitude proportional to RPM
	Hall (based on pulse counting)	M-M. Magnetic field—Hall effect	Automotive ABS, ignition timing
Force (weight)	Foil strain gauge	C-PM. Work done against gravity—electrical resistivity	Load cell is analogous to diaphragm in pressure transducer
Pressure/differential pressure	Capacitive	C-M. Work on diaphragm—AC probe	Built-in over-pressure protection
	Silicon piezo-resistive bridge	C-PM. Work on diaphragm—DC resistive probe	MEMS-based
	Foil strain gauge	C-PM. Work on diaphragm—DC resistive probe	Constantan (copper nickel alloy) foil
Level	Float	C-M. Liquid does work on the float	Level measurement
	Ultrasonic	Ultrasonic, radar, laser beam probes	Level measurement
	Fiber optic in liquid	M-M. Optical probe—photodiode	used for temperature sensing

(continued)

Table 2.1 (continued)

Measurand/input variable	Sensor/transducertechnology	Sensing subsystem mechanism	Applications
Flow	Orifice plate	C. Flow converted to differential pressure	Flow measurement
	Rota meter (turbine)	C-M. Flow to rotation—magnetic	Flow measurement
	Vortex	C-C. DC flow to oscillatory flow—piezoelectric	Sense frequency of oscillations
	Magneto inductive flow	M. Magnetic field probe (generates transverse voltage)	Only for conducting liquids—water, acids
Thermal quantities			
Temperature	Thermocouple	C. Peltier effect generates voltage	Process monitoring and control
	RTDs	PM. Electrical resistivity	
	Thermistor	PM. Electrical resistivity	
	IR pyrometer (non contact type)	D-PM. Lens—bolometer	Measures temperature rise at focal plane
Optical			
Binary light intensity	Photoconductors	PM. Resistivity modulation	Outdoor, building lighting controls
Light intensity	Phototransistors	M. Reverse bias photodiode	Stored optical data, CD, DVD, CD-ROM
Optical barcode	Bar code scanners	M-D-M. Scanned laser—filter-pin diode	Product identification, checkout counters. Market includes assembled system
Optical image (linear arrays)	Document scanners	M, D-C. Scanner, lens—photodiode arrays	Office copiers, image digitization. Market includes diodes only
Acoustic and vibration quantities			
Mechanical vibration in solids	Piezoelectric vibration sensors	C. Mechanical to electrical	Automotive knock sensors, vibration control and measurement
Medical diagnosis	Ultrasonic imaging	M-C. Ultrasonic beams used as probes—detectors	Piezoelectric arrays for density/structural information

(continued)

Table 2.1 (continued)

Measurand/input variable	Sensor/transducertechnology	Sensing subsystem mechanism	Applications
Electrical and magnetic quantities			
Electrical current	Shunts for high currents, catheter electrodes	C. Current to voltage	Power measurement and control, biomedical measurements
Data stored on magnetic media	Magneto inductive	M-C. Changing flux induces electrical voltage	Computers and office equipment
Qualities of materials, liquids, the environment			
Rain on windshield	Rain sensor	M-C. Infrared probe	Windshield wiper control
Metal particles in oil	Oil quality	M. Electrical conductivity	Automobile maintenance
Chemical quantities in liquids and gases			
Humidity	Humidity-capacitive/resistive	Dielectric and dimensional change	Process control
pH in aqueous solutions	pH Electrode	D-C. Selective membrane/ potentiometric	Process control
Gas sensor(dissolved/ exhaust air)	Tin oxide-based (taguchi gas sensor)/ semiconductor/electrochemical	PM. Electrical resistivity C-M selective membrane amperometric	Safety Quality control
Biosensors			
Blood glucose	Glucose biosensor	Chemical reaction amperometric	Control of diabetes (disposable and non disposable)
Medical	Antibody/DNA/Chips	Chemical reactions fluorescent tags	Biological information

Energy conversion (*C*), energy dispersion (*D*), energy modulation (*M*), and property modulation (*P*)

resolution ADCs. Linearization and calibration is then performed digitally, thereby reducing cost and complexity [3, 4].

Commonly sensed measurand:

The primary measurand sensed by temperature sensor is temperature. Only one second measurand has been detected with any practicality—fluid velocity. Temperature sensors based on different principles have been developed. They can be broadly classified into following groups, i.e., electrical, mechanical, and radiation type.

Classification of temperature sensor:

- Contact type:
  - Mechanical

Bimetal
Thermal expansion
magnetic
  - Electrical

Thermocouple
Resistance
Thermistor
Semiconductor
- Noncontact
  - Radiation Pyrometer
  - Optical Pyrometer

Temperature Scales:
Two fixed points are defined for temperature scales.

- ICE POINT: The lower fixed point or ice point is the temperature of ice, prepared from distilled water when melting under a pressure of 760 mm of Hg.
- STEAM POINT: The upper fixed point or steam point is the temperature of steam from pure distilled water, boiling under a pressure of 760 mm of Hg.

The temperature interval between the ice point and steam point is known as the "FUNDAMENTAL INTERVAL".

Different temperature scales are as below:

- Centrigrade or Celcius scale (°C):
  It was introduced in about 1740 by a Swedish astronomer and professor Celcius, and is mostly used in European countries. It has ice point at 0 °C and steam point at 100 °C. This scale depends upon the selection of the working substance used.
- Fahrenheit scale (°F):
  It was introduced in about 1665 by a German philosopher Fahrenheit, and is used in most English-speaking countries. It has ice point at 32 °F and steam

point at 212 °F. The zero point or starting point, 0 °F, represents temperature of certain salt-ice mixture.

$$°C = 5/9(°F - 32)$$

- Kelvin or Absolute scale (K):
  It was introduced in 1848 by Lord Kelvin. It has ice point at 273.16 K and steam point at 373.16 K.

$$K = °C + 273.16$$

- Rankine scale (°R′):
  It has ice point at 491.69 °R′ and steam point at 671.69 °R′. It has 180 degrees between ice point and steam point just as on Fahrenheit scale, therefore it is also called as "absolute Fahrenheit scale".

$$°R' = °F + 459.69$$

- REAUMUR SCALE (°R):
  It is often used in alcohol industries. It has 0 °R as the ice point and 80 °R as the steam point. It was introduced in 1731 and used in few European countries.

An international practical temperature scale has been set up in 1948, in order to define the temperature intervals on the basis of equilibrium of various substances. From −310 °F (−190 °C) to 1,220 °F (660 °C) temperature scale is defined by the resistance of a platinum resistance thermometer. From 1,220 °F (660 °C) to 1,945 °F (1,063 °C) temperature scales are defined by platinum versus platinum–rhodium thermocouple. Above 1,945 °F (1,063 °C) the scale is defined by a formula for the radiation of a black body. Figure 2.4 shows the various types of temperature scale.

The International Practical Temperature Scale (IPTS) defines six primary fixed point for reference temperatures in terms of:

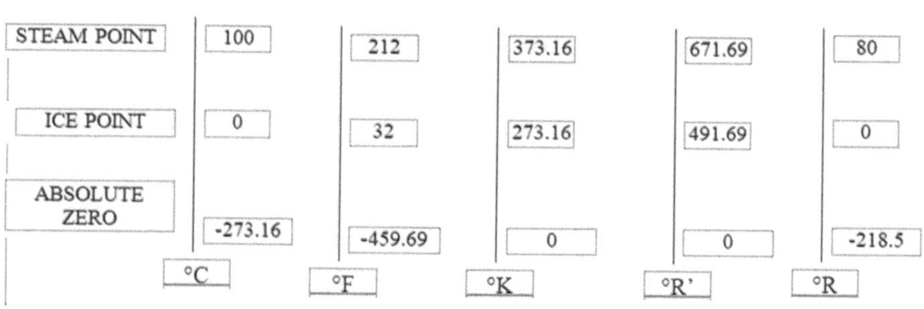

Fig. 2.4 Temperature scale

- the triple point of equilibrium hydrogen −259.34 °C
- the boiling point of oxygen −182.962 °C
- the boiling point of water 100.0 °C
- the freezing point of zinc 419.58 °C
- the freezing point of silver 961.93 °C
- the freezing point of gold 1,064.43 °C

(all at standard atmospheric pressure)

Resistance Temperature Detector (RTD):
The electrical resistance of many metals changes with temperature. RTD is resistive temperature sensor, mostly made up of Platinum. It consists of the carrier, the measuring winding and the connecting wires. A two-wire circuit with Pt 100 RTD's is appropriate for simple measuring applications. Commercially available control, measuring and evaluating equipment is considered for the Pt 100 connection with a three wire circuit. In laboratories and mainly difficult applications in industrial measuring systems four-wire circuit with Pt 100 is used. In Fig. 2.5 types of RTD circuit is shown.

Thermocouple:
Thermocouples are small, rugged, relatively inexpensive, and operate over the widest range of all temperature sensors. They are especially useful for making measurements at extremely high temperatures (up to 2,300 °C) in hostile environments. They produce only millivolts of output, however, and require precision amplification for further processing. They also require cold-junction compensation (CJC) techniques. Principle of thermocouple is explained in Fig. 2.6.

The most common metals used are Iron, Platinum, Rhodium, Rhenium, Tungsten, Copper, Alumel (composed of Nickel and Aluminum), Chromel (composed of Nickel and Chromium), and Constantan (composed of Copper and Nickel) as shown in Table 2.2 [5].

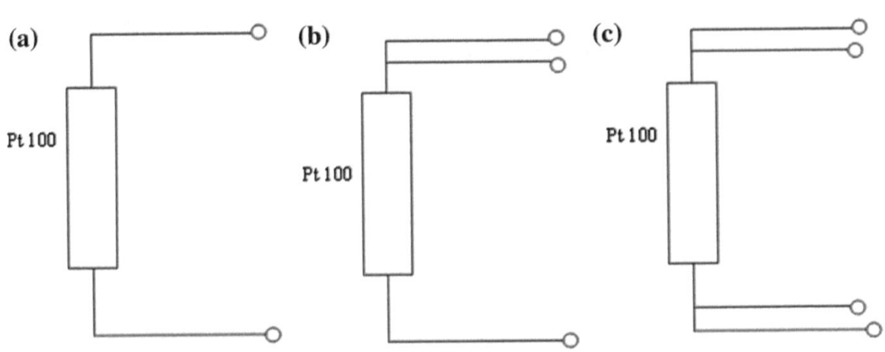

Fig. 2.5 Types of RTD circuit. **a** Two-wire, **b** Three-wire, **c** Four-wire circuit

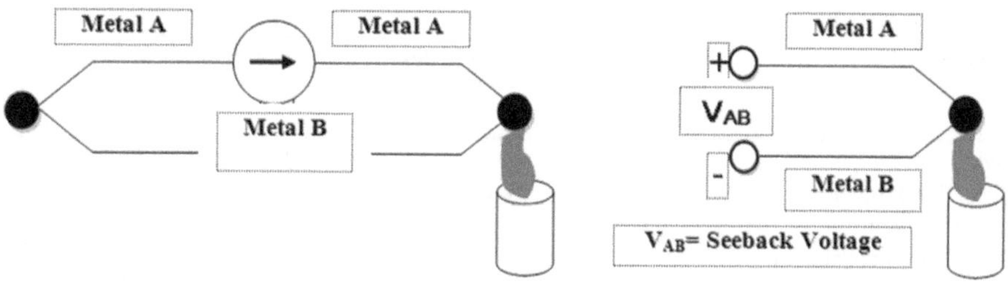

Fig. 2.6 Principle of thermocouple

Table 2.2 Types of thermocouples

	Typical useful range (°C)	ANSI configuration
Platinum(6 %)/Rhodium–Platinum (30 %)/Rhodium	38–1,800	B
Tungsten (5 %)/Rhenium–Tungsten (26 %)/Rhenium	0–2,300	C
Chromel–Constantan	0–982	E
Iron–Constantan	0–760	J
Chromel–Alumel	−184–1,260	K
Platinum (13 %)/Rhodium–Platinum	0–1,593	R
Platinum (10 %)/Rhodium–Platinum	0–1,538	S
Copper–Constantan	−184–400	T

Thermistor:
A thermistor is made of a semiconductor material that exhibits a predictable and repeatable change in resistance as temperature is changed. Unlike a metal, the molecular structure of a semiconductor is such that increasing its temperature reduces its resistance. Thermistors are low cost temperature-sensitive resistors and are constructed of solid semiconductor materials which exhibit a positive or negative temperature coefficient (PTC or NTC).

Pyrometers:
Pyrometers are used for indirect (non-contact) temperature measurements. Pyrometers are used at standard measuring temperatures from −150 °F to 6,300 °F (−100 to 3,500 °C) and up to 9,000 °F (5,000 °C) in special cases.

Basic principle of the pyrometer is the thermal radiation (infrared to visible range) of a measured object filtered optically and concentrated on a radiation receiver. Its electrical reaction consists of a change in the resistance, voltage, or current of the radiation receiver induced directly or indirectly via a temperature increase depending on the principle used. The electrical change is amplified, measured, and processed further.

Pyrometers can be used, if:

- the surface temperature is to be determined,
- the medium to be measured is visually accessible (window, openings, optical fiber).

Table 2.3 Various types of temperature sensors

Sensory type	Limits of application (°C)	Advantages	Disadvantages
Thermocouple			Minimum span of 40 °C
Type E: Chromel-constantan	−100 to 1000	Good reproducibility	Temperature versus emf not exactly linear
Type J: iron-constantan	0 to 750	Wide range	Drift over time
Type K: chromel-Alumel	0 to 1,250		Low emf corrupted by noise
Type T: copper-constantan	−160 to 400		
RTD	−200 to 650	Good accuracy Small span possible Linearity	Self-heating Less physically rugged Self-heating error
Thermistor	−40 to 150	Good accuracy Little drift	Highly nonlinear Only small span Less physically rugged Drift
Bimetallic	–	Low cost Physically rugged	Local display
Filled system	−200 to 800	Simple and low cost No hazards	Not high temperatures Sensitive to external pressure

The freezing points of certain other metals are also used as *secondary fixed points* to provide additional reference points during calibration procedures [4]. Various types of temperature sensors with advantages and disadvantage summarize in Table 2.3.

Temperature is measured on mainly three scales.

- Degree Fahrenheit (°F) (used in U.S. and many other English-speaking countries)
- Degree Celsius (°C) (used in scientific measurements and industrial applications)
- Kelvin (K) (extremely low temperature).

2.4.2 Pressure Sensor

Pressure is the force per unit area that a fluid exerts on its surroundings. Measurement of pressure inside a pipeline or a container in an industrial environment is a challenging task, keeping in mind that pressure may be very high, or very low (vacuum); the medium may be liquid, or gaseous. Pressure sensors operate on the basis of the same principle: the detection of a physical force which arises due to pressure [4, 6].

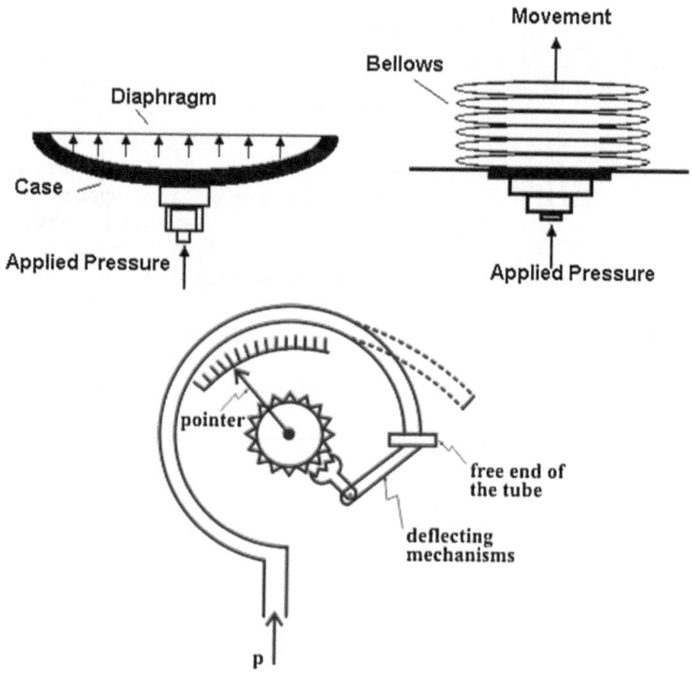

Fig. 2.7 Various types of pressure sensors (diaphragm, bellows, bourdon tube)

Primary pressure sensor
Pressure measurement mainly carried out by using elastic elements: diaphragms, bellows, and Bourdon tubes, are as shown in Fig. 2.7. These elastic elements change their shape with applied pressure and the change of shape can be measured using suitable deflection [7]. Classification of pressure sensor is summarized in Table 2.4.

Units

1 psi (pound per square inch) = 1 lb/in.^{2} = 6.9 kPa;
1 Pa = 1 N/1 m^2
1 bar = 105 Pa.

Gauge pressure: Gauge Pressure is represented by defining atmospheric pressure as zero. Pressure higher than atmospheric pressure is positive pressure, while lower than atmospheric pressure is negative pressure. The difference between the absolute pressure and the local atmospheric pressure is called gauge pressure. Representation of pressure scales is shown in Fig. 2.8.
Absolute pressure: This is the difference between the pressure of the fluid and the absolute zero of pressure. Pressure is represented by defining absolute vacuum as zero. Gauge pressure is generally used. Absolute pressure is used for scientific

Table 2.4 Types of pressure sensors

Sensory type	Limits of application	Advantages	Disadvantages
Bourdon (C) Spiral Helical	Up to 100 MPa	Low cost with reasonable accuracy Wide limits of application	Hysteresis Affected by shock and vibration
Bellows	Typically vacuum to 500 kPa	Low cost Differential pressure	Smaller pressure range of application Temperature compensation needed
Diaphragm	Up to 60 kPa	Very small span possible	Usually limited to low pressures
Resistive/strain gauge	Up to 30 kPa	Large range of pressures	Sensitive
Piezoelectric	–	Fast dynamics	Sensitive to temperature changes

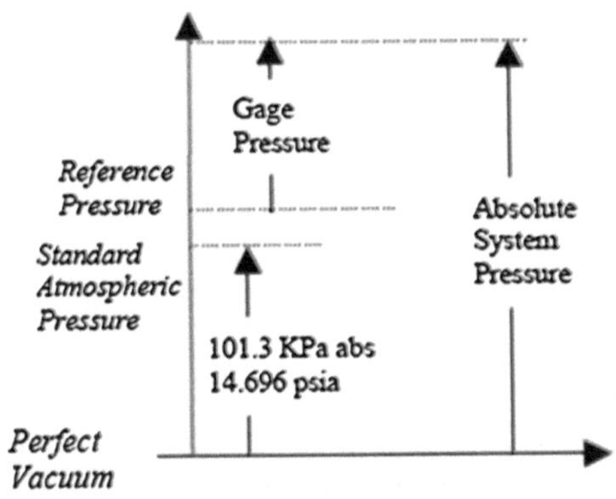

Fig. 2.8 Pressure scales

calculations and is distinguished from gauge pressure by putting "abs" after the unit.

$$\text{Absolute pressure} = \text{Gauge pressure} + \text{Atmospheric pressure}$$

Differential pressure: This term is used to describe the difference between two absolute pressure values, such as the pressures at two different points within the same fluid (often between the two sides of a flow restrictor in a system measuring volume flow rate).

In many cases, the absolute pressure is not the quantity of major interest in describing the pressure. The atmosphere of gas that surrounds the earth exerts a pressure, because of its weight, at the surface of the earth of approximately 14.7 psi. If a closed vessel at the earth's surface contained a gas at an absolute pressure of 14.7 psi, then it would exert no effective pressure on the walls of the container because the atmospheric gas exerts the same pressure from outside. In this case, it is more appropriate to describe the pressure in a relative sense that is compared to atmospheric pressure. This is called gauge pressure and is given by:

$$P_g = P_{abs} - P_{at}$$

where,
P_g gauge pressure,
P_{abs} absolute pressure,
P_{at} atmospheric pressure.

Methods for measuring pressure:

1. Direct-measuring pressure instruments determine the pressure from the basic equation:

$$P = F/A$$

Types of Direct-measuring pressure instruments

- Liquid column manometers (U-tube, inclined-tube, Multiple liquid, Float-type)
- Pressure balances with liquid separation (immersed segment, immersed cylinder, and the cylindrical)
- Piston-type pressure measuring instruments (dead weight tester).

2. Indirect-measuring pressure instruments use the deflection of a flexible material or an electrical, optical, or chemical effect to determine the measured pressure. Measuring converters are instruments which convert the pressure acting on them into an output which is generally an electric or pneumatic signal. This output is a function of the input pressure and can be either digital or analog [8].

- Diaphragm
- Strain guage
- Piezo resistive element
- Inductive
- Capacitive
- Hall effect

Many other types of instruments exist for measuring low and high pressures in addition to those based on electrical pressure measuring systems presented above. They include:

- McLeod compression gauges
- Pirani vacuum gauges based on the principle of thermal conductance
- Ionization pressure gauges
- Friction pressure gauges which use the internal friction of gases as the basis for pressure measurements.

Table 2.5 **a** Pressure conversion table, **b** Torque (moment of force), **c** Pressure, **d** Force

a[a]

	mmHg	mmH_2O	Kgf/cm^2	Atm	Bar	Psi	Pa
mmHg	1	13.60	1.360×10^{-3}	1.316×10^{-3}	1.333×10^{-3}	1.933×10^{-2}	133.3
mmH_2O	7.356×10^{-2}	1	1×10^{-4}	0.968×10^{-4}	0.981×10^{-4}	1.422×10^{-3}	9.8067
Kgf/cm^2	735.6	10,000	1	0.968	0.981	14.22	98,067
Atm	760	10,332	1.033	1	1.013	14.71	101,325
Bar	750.1	10,197	1.020	0.987	1	14.50	100,000
Psi	51.72	703.1	0.070	0.068	0.069	1	6,895
Pa	7.501×10^{-3}	0.102	1.020×10^{-5}	9.869×10^{-6}	1×10^{-5}	1.45×10^{-4}	1

b[b]

	N m	kg_f m	pdl ft	lb_f ft
N m	1	0.1020	23.73	0.7376
kg_f m	9.807	1	232.7	7.233
pdl ft	0.04214	4.297×10^{-3}	1	0.03108
lb_f ft	1.356	0.1383	32.17	1

c[c]

	mbar	Bar	N/m^2	lb/in^2	In Hg	atm
mbar	1	10^{-3}	100	0.01450	0.02953	9.869×10^{-4}
Bar	1000	1	10^5	14.50	29.53	0.9869
N/m^2	0.01	10^{-5}	1	1.450×10^{-4}	2.953×10^{-4}	9.869×10^{-6}
lb/in^2	68.95	0.06895	6,895	1	2.036	0.06805
In Hg	33.86	0.03386	3,386	0.4912	1	0.03342
atm	1013	1.013	1.013×10^5	14.70	29.92	1

d[d]

	N	kg_f	kN	pdl	lb_f	UK ton_f
N	1	0.1020	10^{-3}	7.233	0.2248	1.004×10^{-4}
kg_f	9.807	1	9.807×10^{-3}	70.93	2.2046	9.842×10^{-4}
kN	1000	102.0	1	7233	224.8	0.1004
Pdl	0.1383	0.0141	1.383×10^{-4}	1	0.0311	1.388×10^{-5}
lb_f	4.448	0.4536	4.448×10^{-3}	32.174	1	4.464×10^{-4}
UK ton_f	9,964	1,016	9.964	72,070	2,240	1

[a] Torr: unit used to represent absolute pressure close to absolute vacuum. *Example* 1 Torr = 1 mmHg abs, 10^{-3} Torr = 0.001 mmHg abs

[b] *SI unit* N m, *Metric unit* kg_f m, *Imperial units* pdl ft, lb_f ft

[c] *SI unit* mbar, bar, N/m^2 , *Imperial units* lb/in^2 , in Hg, atm

[d] *SI units* N, kN, *Metric unit* kg_f, *Imperial units* pdl, lb_f, UK ton_f

Various types of pressure unit and conversion to another unit are shown in Table 2.5a–d [16].

2.4.3 *Flow Sensor*

Flow rate is related to pressure by causing the flowing fluid to pass through some form of restriction in the transport pipe, which creates a momentary loss of pressure. This pressure differential is related mathematically to the flow rate [8]. Flow is usually monitored in open channels (gravity flow) and closed conduits (pressurized flow in completely closed pipes).

Flow Q is the volume flowing through a mathematically precise determined cross section over a certain time unit.

$$Q = \text{Volume}/\text{Time} = V/t$$

A flow sensor is a device that measures the flow volume of water, pure water, compressed air, gas, drugs, steam, and similar liquids or gases. Flow sensors are of various types that each operates with different detection principles [6].

Flow sensors are used in many examining and control applications, to measure both air and liquid flows [7]. There are many ways of defining flow (mass flow, volume flow, laminar flow, turbulent flow). Usually the amount of a substance flowing (mass flow) is the most important, and if the fluid's density is constant, a volume flow measurement is a useful substitute that is generally easier to perform [9]. Various types of flow sensors are shown in Fig. 2.9 and also summarized in Table 2.6.

Flow measurements include measuring of flow rate of solids, liquids, and gases. There are two basic ways of measuring flow; one on volumetric basis and the other on weight basis. Concrete materials are measured in terms of either weight per unit time or mass per unit time. Very rarely hard quantity is measured in terms of volume. Liquids are measured either in volume rate or in weight rate. Gases are normally measured in volume rate.

For measurement of flow various methods are used, i.e., thermal anemometers, differential pressure measurement systems, and vortex shedding sensors. Methods used for measuring liquid flow are: differential pressure measurement systems, vortex shedding sensors, positive displacement flow sensors, turbine-based flow sensors, magnetic flow sensors, and ultrasonic flow sensors [9].

Flow Sensor:

- Electrical
  - Electro magnetic sensors
- Mechanical
  - Turbine meters
  - Rotameters
  - Karman vortex meters
  - Coriolis meters
- Others
  - Ultrasonic meters
  - Thermal meters.

Units

The units used to describe the flow measured can be of several types depending on how the specific process needs the information.

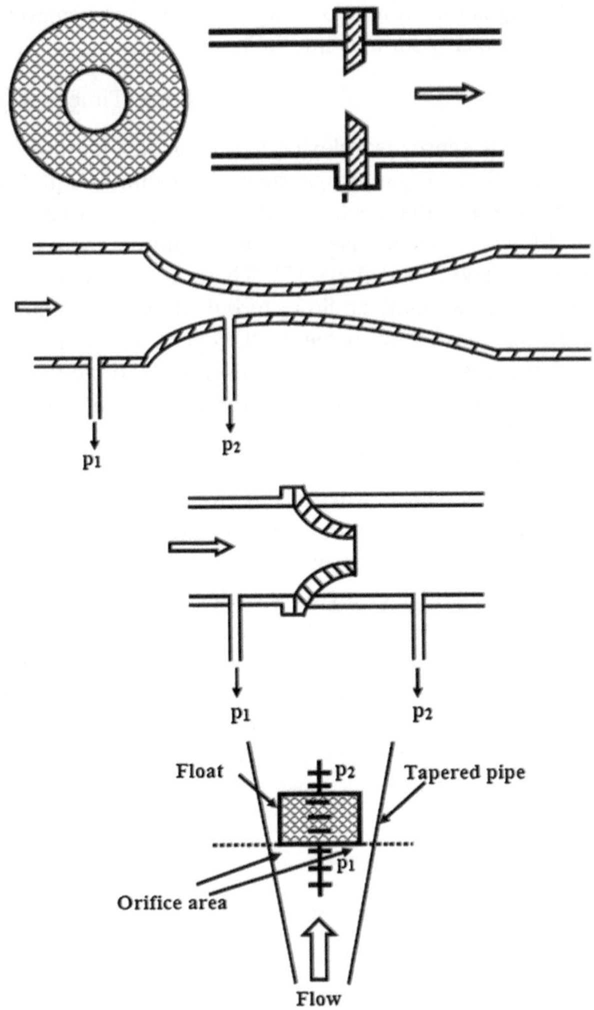

Fig. 2.9 Type of flow sensor: orifice plate, venturi meter, flow nozzle, rotameter

Solids	Tonnes/h, kg/min etc.
Liquids	Tonnes/h, kg/min, L/h, L/min, m^3/h etc.
Gases	Std m^3/h, N m^3/h etc.
Steam	Tonnes/h, kg/min etc.

Steam density at different temperatures and pressures vary. Hence the measurement is converted into weight rate of water which is used to produce steam at the point of measurement. Flow rate measurement and conversion units are shown in Table 2.7a and b.

Table 2.6 Types of flow sensors

Sensory type	Advantages	Disadvantages
Orifice	Low cost Extensive industrial practice	High pressure loss Plugging with slurries
Venture	Lower pressure loss than orifice Slurries do not plug	High cost Line under 15 cm
Flow nozzle	Good for slurry service Intermediate pressure loss	Higher cost than orifice plate Limited pipe sizes
Elbow meter	Low pressure loss	Very poor accuracy
Turbine	Wide range ability Good accuracy	High cost Strainer needed Especially for slurries
Positive displacement	High range ability Good accuracy	High pressure drop Damaged by flow surge or solids

Table 2.7 **(a)** Mass flow rate, **(b)** Volume flow rate

a[a]

	g/s	kg/h	Tonne/d	lb/s	lb/h	Ton/d
g/scq	1	3.6	0.08640	2.205×10^{-3}	7.937	0.08503
kg/h	0.2778	1	0.02400	6.124×10^{-4}	2.205	0.02362
Tone/d	11.57	41.67	1	0.02551	91.86	0.9842
lb/s	453.6	1633	39.19	1	3600	38.57
lb/h	0.1260	0.4536	0.01089	2.788×10^{-4}	1	0.01071
Ton/d	11.76	42.34	1.016	0.02593	93.33	1

b[b]

	l/h	ml/s	m^3/s	Gal/h	ft^3/s	ft^3/h
l/h	1	0.2778	2.778×10^{-7}	0.2200	9.810×10^{-6}	0.035316
Ml/s	3.6×10^6	1	10^{-6}	0.7979	3.532×10^{-5}	0.12714
m^3/s	3.6	10^6	1	7.919×10^5	35.31	1.271×10^5
Gal/h	4.546	1.263	1.263×10^{-6}	1	4.460×10^{-5}	0.16056
ft^3/s	1.019×10^5	2.832×10^4	0.02832	2.242×10^4	1	3600
ft^3/h	28.316	7.8653	7.865×10^{-6}	6.2282	2.778×10^{-4}	1

[a] *SI unit* g/s, *Metric units* kg/h, tonne/d, *Imperial units* lb/s, lb/h, ton/d
[b] *SI unit* m^3 /s, *Metric units* l/h, ml/s, *Imperial units* gal/h, ft^3 /s, ft^3 /h

2.4.4 Level Sensor

In industry, usually vast quantities of liquids such as water, solvents, chemicals, etc., are used in a number of industrial processes. Liquid level measurements are made to ascertain the quality of liquid held in a container or vessel. The liquid level affects both pressure and rate of flow in and out of the container and therefore its measurement and/or control becomes quite important in a variety of processes encountered in modern manufacturing plants. Level sensing is very much related to flow sensing. The common application for level sensing is tank level

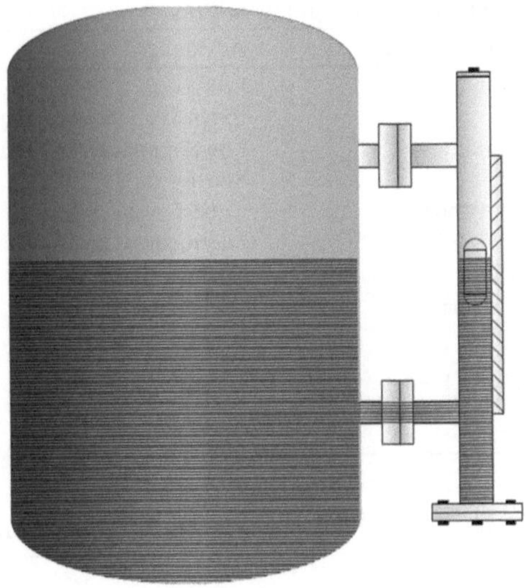

Fig. 2.10 Magnetic float type level indicator

Table 2.8 Types of level sensors level sensors

Sensory type	Limits of application	Advantages	Disadvantages
Float	Up to 1 m	Can be used for switches	Cannot be used with sticky fluids which coat the float
Displacement	0.3–3 m	Good accuracy	Limited range Cost of external mounting for high pressures
Differential pressure	No upper limit	Good accuracy Large range Applicable to slurries with use of sealed lines	Assumes constant density Sealed lines sensitive to temperature
Capacitance	Up to 30 m	Applicable for slurries Level switch for many difficult fluids	Affected by density variations

measurement and control operations. Figure 2.10 shows the level sensing in tank measurement using magnetic float-type level indicator [17].

Level measurements are classified into two groups: direct methods and indirect methods. A number of level sensing technologies are currently available, including manometer, float method, ultrasonic, fluid pressure, RF capacitance, magneto restrictive-based, and radar measurement systems. Various types of flow sensors are summarizes in Table 2.8.

Classification of Level sensors:

- Mechanical sensors
  - Float methods

- Buoyancy method
- Vibrating level systems

• Hydrostatic pressure methods

- Differential pressure level detectors
- Bubbler systems

• Electrical methods

- Conductivity probes
- Capacitance probes
- Optical level switches
- Ultrasonic level detectors
- Microwave level systems
- Nuclear level systems

Units:

It is expressed in terms of pressure the column exerts over a datum level or the length of the liquid column.

2.4.5 Electrochemical Sensor

Electrochemical sensors operate by reacting with the gas of interest and producing an electrical signal proportional to the gas concentration. A typical electrochemical sensor consists of a *sensing electrode* (or working electrode), and a *counter electrode* separated by a thin layer of electrolyte. A schematic of capacitive sensor is shown in Fig. 2.11.

Electrochemical sensors require very little power to operate. Their power consumption is the lowest among all sensor types available for gas monitoring. So, the sensors are widely used in portable instruments that contain multiple sensors. They are the most popular sensors in confined space applications. A sensor's life expectancy is highly dependent on the environmental contaminants, temperature, and humidity to which it is exposed.

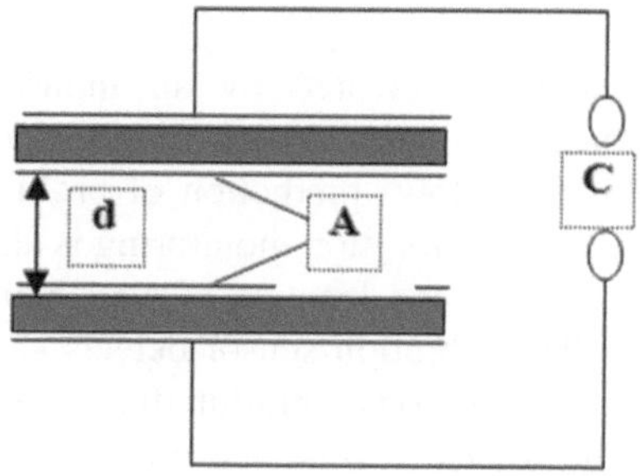

Fig. 2.11 Schematic capacitive sensor

Capacitive sensors:
Capacitive sensor is one of the modern developments in sensor transducer technology [10]. Capacitive sensors directly sense a variety of things—motion, chemical composition, electric field and, indirectly, sense many other variables which can be converted into motion or dielectric constant, such as pressure, acceleration, fluid level, and fluid composition. They are built with conductive sensing electrodes in a dielectric, with excitation voltages and detection circuits which turn a capacitance variation into a voltage, frequency, or pulse width variation. The range of application of capacitive sensors is amazing [10].

Capacitive displacement sensors "are non-contact devices capable of high-resolution measurement of the position or change in position of the target". They are also able to measure the thickness or density of non-conductive materials. Capacitive displacement sensors are used in a wide variety of applications including semiconductor processing, assembly of precision equipment such as disk drives, precision thickness measurements, machine tool metrology, and assembly line testing [10].

Application:

- Capacitive technology is displacing Piezo resistance in silicon implementations of accelerometers and pressure sensors, and innovative applications like fingerprint detectors and infrared detectors are appearing on silicon with sensor dimensions in the microns and electrode capacitance of 10 fF, with resolution to 5×10^{-18} F.
- Capacitive sensors in oil refineries measure the percent of water in oil, and sensors in grain storage facilities measure the moisture content of wheat.
- In the home, cost-effective capacitive sensors operate soft-touch dimmer switches and help the home craftsman with wall stud sensors and digital construction levels.
- Laptop computers use capacitive sensors for two-dimensional cursor control and transparent capacitive sensors on computer monitors are found in retail kiosks.
- Many applications like flow (convert flow to displacement), pressure, liquid level, thickness measurement, and many more are there where capacitor sensor is used.

2.4.6 Vibration Sensor

Vibrations created by an industrial machine are a direct indication of the machine's health monitoring programs and recording of the machine's vibration history allows prediction of problems and shut downs a machine before serious damage. Vibration monitoring is also widely used as a diagnostic tool to determine the cause and location of a problem, and how to fix it.

The vibration sensor detects shock intensity caused by sudden knocks or hits and continuous vibration due to faulty ball-bearings on fans and other equipment. The shock levels and monitoring durations can be set for each individual sensor, specific applications and equipment. Despite the advances made in vibration monitoring and analysis equipment, the selection of sensors and the way they are mounted on a machine remain critical factors in determining the success of any monitoring list.

The three parameters detected in place of motion, by vibration monitors are:

1. Displacement
2. Velocity
3. Acceleration

These parameters can be measured by a variety of motion sensors and are mathematically related. Selection of a sensor proportional to displacement, velocity, or acceleration depends on the frequencies of interest and the signal levels concerned.

Transducers involved in vibration measurement:

- LVDT
- Eddy current
- Capacitive
- Hall

Velocity Sensors

Velocity sensors are used for low to medium frequency measurements. They are useful for vibration monitoring and balancing operations on rotating machinery. As compared to accelerometers, velocity sensors have lower sensitivity to high frequency vibrations. The mechanical design of the velocity sensor; an iron core moving within a coil in a limited magnetic field, no clipping of the generated signal occurs, but smooth saturation. Velocity sensor is used for medium to low frequency (1–1,000 Hz) measurements. It acts as a low-pass filter (reduce high frequency signals). Traditional velocity sensors employ an electromagnetic sensor to pick up the velocity signal.

Accelerometers

Accelerometers are widely used to measure tilt, inertial forces, shock, and vibration. They find wide usage in automotive, medical, industrial control, and other applications. Acceleration integrated to velocity can be used for low frequency measurements.

The basic acceleration sensor has a good signal to noise ratio over a wide dynamic range. They are useful for measuring low to very high frequencies and available in a wide variety of general purpose and application-specific designs. Acceleration sensors come in a variety of sorts both regarding their performance as well as the underlying principles used for sensing. On the rough end of the spectrum one finds the sensors meant to be used, e.g., in car as airbag deployment sensors whereas on the other end the very sensitive (micro-g) sensors are found which are intended for use, e.g., for seismic applications. Accelerometers are extremely versatile and widely used for industrial machinery monitoring. Mainly industrial accelerometers manufactured of a Piezo-ceramic material sandwiched between a seismic mass and the structure base. The seismic mass and Piezo-ceramic create a simple mass/spring system with a very high natural frequency. The piezoelectric sensor is versatile, reliable, and the most popular vibration sensor for machinery monitoring. Capacitive accelerometers are popular due to low-cost production, high sensitivity, good dc-response and noise-performance, low power dissipation, and a simple structure.

Displacement Sensor

Displacement sensors measure the distance where an object moves and they can also be used to measure object height and width. There are mainly two types of displacement sensors: Contact type and not contact type. Non-contact sensors are based on various technologies including electric field, electromagnetic field, and light/laser (capacitive type, inductive type). Displacement sensor can provide contactless linear displacement measurement of absolute position in hydraulic and pneumatic cylinders and due to this reason they offer wide application in industries. Displacement sensors are widely used not only to measure the distance of a moving object but also it can be embedded in other sensors or transducer devices to measure pressure or level or flow.

Eddy current probes are non-contact sensors primarily used to measure shaft vibration, shaft/rotor position, and clearance. Also referred to as displacement probes, eddy current probes are typically applied on machines utilizing sleeve/journal bearings. They have excellent frequency response with no lower frequency limit and can also be used to provide a trigger input for phase-related measurements [11].

Different types of mechanical sensors can be considered such as [7]: Potentiometer type (linear, Rotary), Capacitive and inductive type, Variable reluctance type (*LVDT*), Strain gauges. Figure 2.12 shows various types of displacement sensors. Table 2.9 summarizes the displacement sensors comparison by detection methods.

LVDT-Linear Variable Differential Transformer (*Positional Sensor*)

LVDT is linear position sensor, which can measure movements as small as a few millionths of an inch up to several inches, and also capable of measuring positions up to ±20 in. (±0.5 m). Figure 2.13 shows the components of a typical LVDT. The transformer's internal structure consists of a primary winding centered between a pair of identically wound secondary windings, symmetrically spaced about the primary. The coils are wound on a one-piece hollow form of thermally stable glass reinforced polymer, encapsulated against moisture, wrapped in a high permeability magnetic shield, and then secured in cylindrical stainless steel housing. This coil assembly is usually the stationary element of the position sensor. The moving element of an LVDT is a separate tubular armature of magnetically permeable material called the core, which is free to move axially within the coil's hollow bore, and mechanically coupled to the object whose position is being measured. This bore is typically large enough to provide substantial radial clearance between the core and bore, with no physical contact between it and the

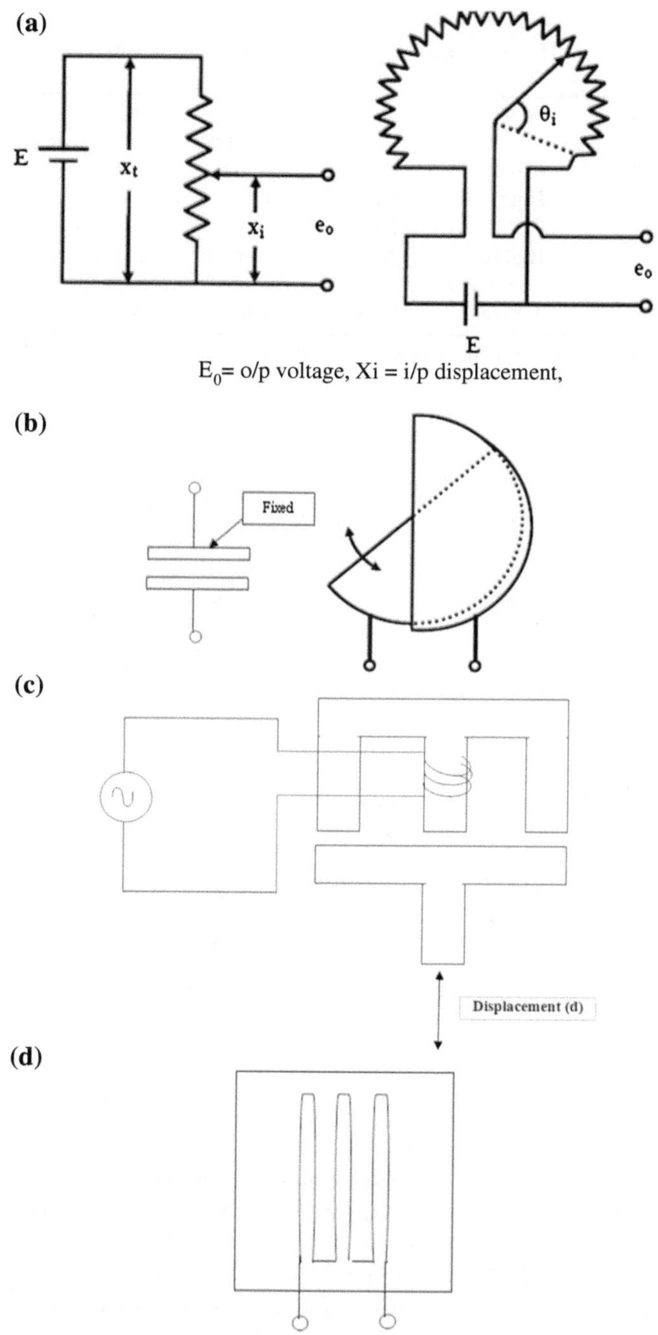

Fig. 2.12 Various types of displacement sensor. **a** Potentiometer (*i*) linear (*ii*) rotary, **b** capacitive type displacement sensors (variable section type and variable area type), **c** inductive displacement, **d** strain guage

Table 2.9 Displacement sensor comparison by detection method

Type	Eddy current	Optical	Ultrasonic wave	Laser focus	Contact
Detectable object	Metal	Most objects	Most objects	Most objects	Solids
Detecting distance	Short	Normal	Long	Short	Short
Accuracy	High	High	Low	High	High
Response speed	Fast	Fast	Slow	Normal	Slow
Environmental resistance	Highly durable	Normal	Normal	Normal	Highly durable
Detecting point	Normal	Small	Large	Small	Small

Units In the region of 0.00001 in. or 0.001 μm minimum

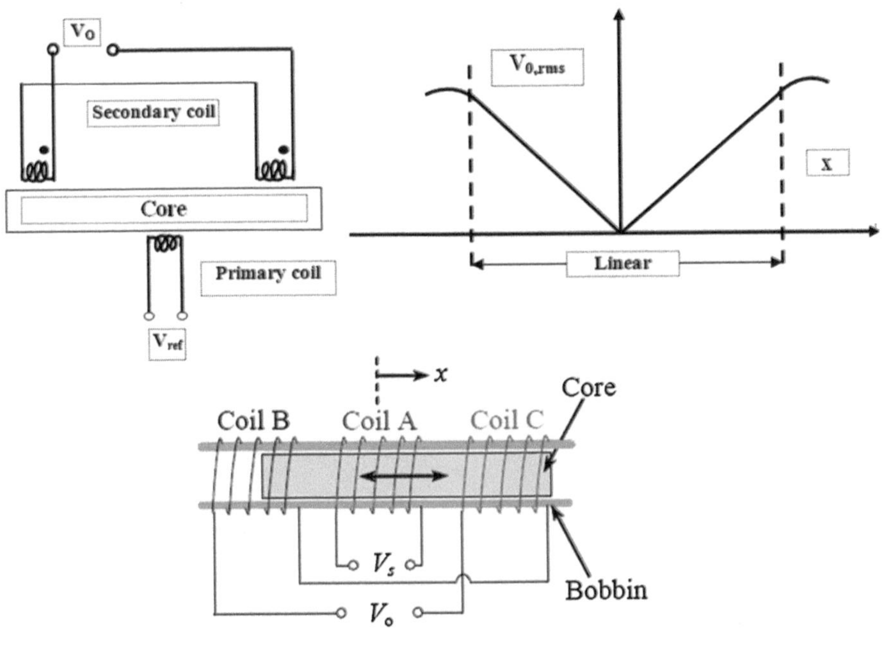

Fig. 2.13 LVDT and its characteristics

coil. In operation, the LVDT's primary winding is energized by alternating current of appropriate amplitude and frequency, known as the primary excitation. Figure 2.13 shows the basic principle of LVDT and its characteristic.

The LVDT's electrical output signal is the differential AC voltage between the two secondary windings, which varies with the axial position of the core within the LVDT coil. Usually this AC output voltage is converted by suitable electronic circuitry to high level DC voltage or current that is more convenient to use.

2.5 Optical Sensor

Optical methods are best-recognized techniques for sensing biochemical analytes. Instrumentation for optical measurements generally consists of a light source, a number of optical components to generate a light beam with specific characteristics and to direct this light to some modulating agent, and a photo-detector for processing the optical signal. The central part of an optical sensor is the modulating component. Figure 2.14 displays the basic principle of optical sensor.

Optical displacement sensors work on the basic principle that the intensity of light decreases with distance. So if the source and detector are fixed, the amount of light reflected from a moving surface will depend on the distance of the moving surface from the fixed ones. Measurement using this principle requires proper calibration since the amount of light received depends upon the reflectivity of the surface, intensity of the source, etc. Yet it can provide a simple method for displacement measurement. Optical fibers are often used to transmit light to and from the measuring zone.

Types of optical sensors:

- Intrinsic
- Extrinsic
- Distributed

Optical sensors are usually based on optical fibers or on planar waveguides. Generally, there are three distinctive methods for quantitative optical sensing at surfaces:

1. The analyte directly affects the optical properties of a waveguide, such as evanescent waves (electromagnetic waves generated in the medium outside the optical waveguide when light is reflected from within) or surface plasmons (resonances induced by an evanescent wave in a thin film deposited on a waveguide surface).
2. An optical fiber is used as a plain transducer to guide light to a remote sample and return light from the sample to the detection system. Changes in the intrinsic optical properties of the medium itself are sensed by an external spectrophotometer.

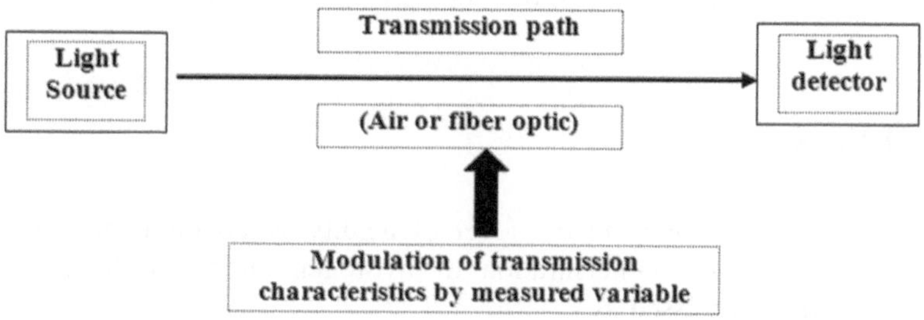

Fig. 2.14 Optical sensor basic principle

3. An indicator or chemical reagent placed inside, or on, a polymeric support near the tip of the optical fiber is used as a mediator to produce an observable optical signal. Typically, conventional techniques, such as absorption spectroscopy and fluorimetry, are employed to measure changes in the optical signal.

2.6 Analytical Measurement Sensor

Many chemical composition measurements may be made indirectly by means of electricity, if those measurements are related to the concentration of ions (electrically charged molecules). Such measurements include:

- pH of an aqueous solution
- Oxygen concentration in air
- Ammonia concentration in air
- Lead concentration in water

Humidity Measurement

Gas or air humidity measurements are becoming more and more important. Constant improvements to the technical processes, higher demands on quality and energy saving require an accurate, stable, and affordable measuring procedure to measure air humidity.

Dew point Measurement

Humidity measurement finds wide applications in different process industries. Moisture in the atmosphere must be controlled below a certain level in many manufacturing processes, e.g., semiconductor devices, optical fibers, etc. Humidity inside an incubator must be controlled at a very precision level. Textiles, papers and cereals must be dried to a standard storage condition in order to prevent the quality deterioration. Figure 2.15 shows the dew point meter illustration.

The humidity can be expressed in different ways:

1. absolute humidity,
2. relative humidity, and
3. dew point [12].

Various types of analytical sensors comparison is shown in Table 2.10.

2.7 pH Sensor

pH is a parameter to indicate the level of acidity or alkalinity in a chemical solution. It defines the concentration of hydrogen atoms in the solution in grams/liter and is expressed as:

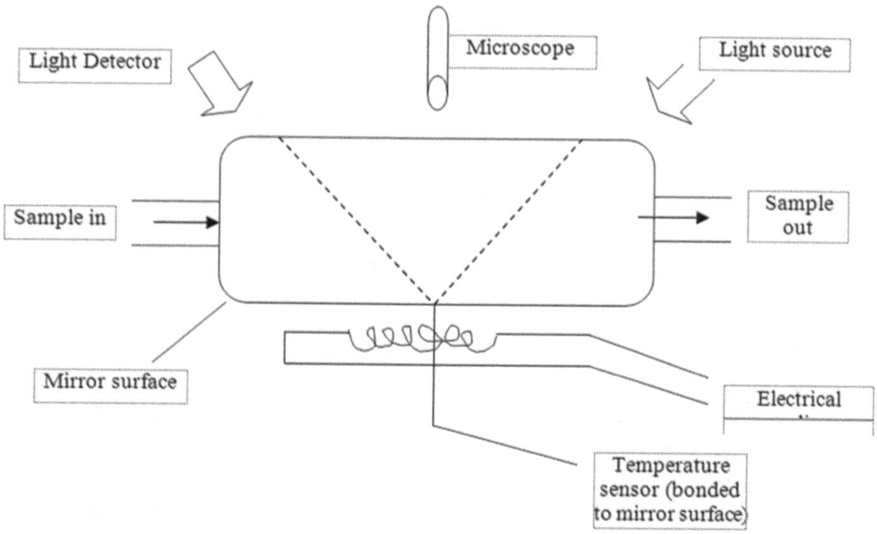

Fig. 2.15 Dew point meter

$$pH = -\log[H+]$$

pH System

A successful pH reading depends upon all components of the system being operational. Problems with any one of the three: electrode, meter, or buffer will give in poor readings.

Electrodes: A pH electrode consists of two half-cells; a demonstrating (Measuring) electrode and a reference electrode. Most applications today use a combination electrode with both half cells in one body. Over 90 % of pH measurement problems are related to the improper use, storage or selection of electrodes. Schematic of pH electrode is shown in Fig. 2.16.

The mathematical expression for pH measurement is:

$$E = E_{\mathrm{m}} - E_{\mathrm{r}}$$

where:

E_{m} the electrode potential of the measuring electrode,
E_{r} the electrode potential of the reference electrode.

Meters: A pH meter is a refined voltmeter capable of reading small millivolt changes from the pH electrode system. The meter is infrequently the source of problems for pH measurements. Today pH meters have temperature compensation (either automatic or manual) to correct for variations in slope caused by changes in temperature. Microprocessor technology has created many new convenience

Table 2.10 Analytical sensors

Hair hygrometer	Psychrometer	Dew point meter	Capacitive humidity sensor
The hair hygrometer is one of the oldest methods used to measure humidity. The length of the hairs changes in accordance with the ambient humidity. This change is mechanically indicated as relative humidity	A temperature probe covered usually with a damp cotton sleeve cools down as a result of evaporation. A second temperature probe measures the ambient temperature. The ambient humidity can be determined from the difference in temperature	A mirror is cooled until it shows condensation after having reached the dew point temperature. The condensation on the mirror is monitored and the dew point is then measured	A condensator changes its capacity in accordance with the ambient humidity
Advantages			
Simple to use measuring engineering with low installation costs Low cost applications	If used with care a very accurate measurement of 2 to 3 %RH is possible	Wide measuring range Highly accurate	Affordable, quick-action and accurate measurement (up to ±1 %RH) Wide measuring range (0–100%RH, −40 to +180 °C) Long-term stability Small, portable measuring instruments
Disadvantages			
High maintenance costs Frequent regeneration of the hairs Can be used only from 15% to 85 %RH and up to max. 50 °C Highly inaccurate, not definable Slow measurements	Cannot be used for multipoint measurements Time-consuming handling (must be moistened with distilled water before nearly every measurement) Before every important measurement, the temperature must be adapted to the ambient temperature and the sleeve should be changed	Time-consuming, expensive method Not battery-operated Heavy (non-portable measuring instrument) Highly accurate temperature measurement required Slow adaptation time Large bench-top instruments	Capacitive sensor has been tested worldwide and has established itself in industrial measurement engineering

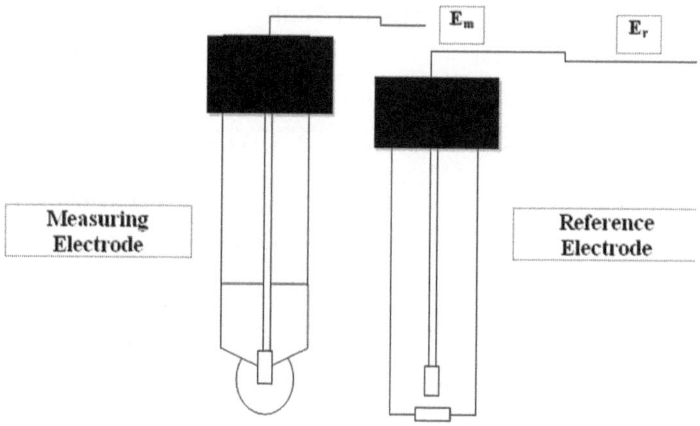

Fig. 2.16 pH Electrode

features for pH measurement; auto-buffer recognition, calculated slope and % efficiency, log tables for concentration of ions and more.

Buffers: These solutions of known pH value allow the user to adjust the system to read accurate measurements. For best accuracy:

- Standardization should be performed with fresh buffer solutions.
- Buffer used should frame the range of pH for the samples being tested.
- Buffers should be at the same temperature as the samples (For example: if all your samples are at 50 °C, warm your buffers to 50 °C using a beaker in a warm bath.). Buffer values are dependent upon temperature. In Fig. 2.17 displays the pH scale group with various types of examples, i.e., Milk pH is near to between 6 and 7pH.

Application of pH measurement:

- Jam and jelly manufacturing
- Laundries
- Meat and Fish Processing
- Pharmaceuticals
- Printing
- Sewage
- Swimming pool
- Dyeing
- Electrical equipment

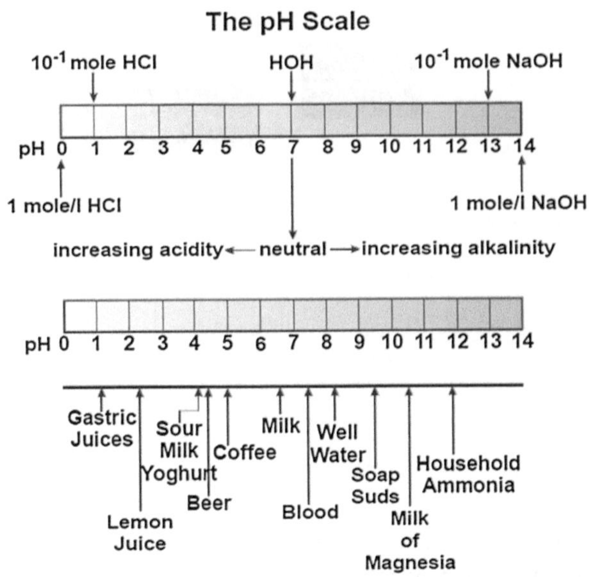

Range	pH	H + concentration(mol/L)	OH - concentration(mol/L)
	0	1	0.00000000000001
	1	0.1	0.0000000000001
	2	0.01	0.000000000001
Acid	3	0.001	0.00000000001
	4	0.0001	0.0000000001
	5	0.00001	0.000000001
	6	0.000001	0.00000001
neutral	7	0.0000001	0.0000001
	8	0.00000001	0.000001
	9	0.000000001	0.00001
	10	0.0000000001	0.0001
alkaline	11	0.00000000001	0.001
	12	0.000000000001	0.01
	13	0.0000000000001	0.1
	14	0.00000000000001	1

Fig. 2.17 pH Scale

2.8 Piezoelectric Sensor

Piezoelectric sensors have proven to be a versatile tool for the measurement of various processes. Piezoelectric substance generates an electric charge when subjected to a sudden change in stress, such as sudden blow to the crystal. It is a mechanical sensor but responds by producing an electrical output only if there is a change in the applied force. The single disadvantage of piezoelectric sensors is that they cannot be used for true static measurements.

Piezoelectric sensors offer a unique set of capabilities that cannot be found in other sensing principles. It contains the inherent temperature stability, the

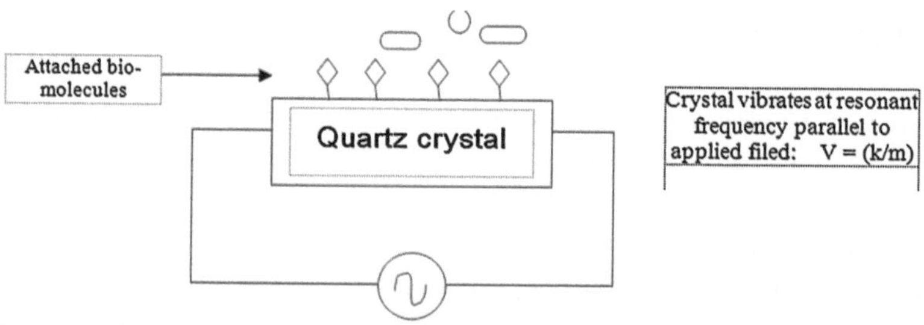

Fig. 2.18 Piezo electric sensor

amplitude range, and the signal quality make it very interesting, and in particular where no static information is needed.

Piezoelectric sensors are electromechanical systems that react on compression; the sensing elements show almost zero deflection. This is the reason why piezoelectric sensors are so rugged, have an extremely high natural frequency and an excellent linearity over a wide amplitude range. Additionally, piezoelectric technology is insensible to electromagnetic fields and radiation, enabling measurements under harsh conditions. Figure 2.18 shows the Piezo electric sensor used in bimolecules detection in analyte.

Advantages of piezoelectric sensors are:

- Reliable, robust, and compact
- Low energy consumption
- Active signal producing component—no powering needed
- Extremely high temperature range
- Linearity
- Time stable and long lasting
- High frequencies
- Bidirectional electromechanical conversion.

Types of Piezo sensors:

- Accelerometers
- Force sensors
- Shock sensors
- Strain and surface deformation sensors
- Sensors for ballistic testing devices
- Bone microphones
- Microphones.

2.9 Photo-conductive Sensor

It is a recent development in solid-state electronics. It is a special material when exposed to light; alter their internal electrical resistance in proportion to the amount of light falling on them [13].

Commonly sensed measurands:

The primary measurands sensed by the photoconductive sensor is light. There are numerous secondary measurands like position, sound, motion, temperature, flow, and force. Figure 2.19 shows the schematic of photoconductive sensor.

Phototransistor:

A phototransistor is a normal transistor in which the envelope enclosing the junction is transparent to allow light to fall on the base emitter junction.

Photodiode:

Photodiodes are PN junction diodes designed specifically to harness the photoelectric effect. There are many useful applications of this phenomenon, such as solar cells, light detection/emission, and thus signal receiving/sending. As of their fast response time, photodiodes are used as film sound track readers. They can also be used as detectors of modulated light in optical communication systems and also in switching circuits. Photodiodes may be biased and operated in two basic modes: photovoltaic and photoconductive.

The advantages of phototransistor/diode:

- low power consumption,
- small size,
- immediate operation on switching on,
- low voltage operation and long life.

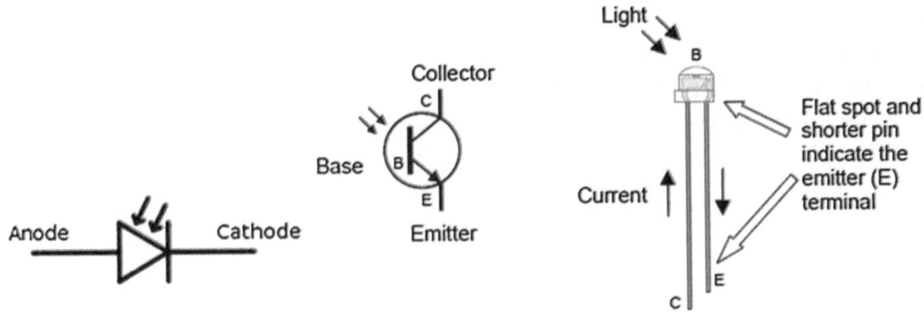

Fig. 2.19 Photoconductive sensor (photodiode/phototransistor)

2.10 MEMS and Fabrication of MEMS Sensor

Micro-Electro-Mechanical Systems (MEMS) is the integration of mechanical elements, sensors, actuators, and electronics on a common silicon substrate through micro-fabrication technology. Electronics are fabricated using integrated circuit (IC) process and the micromechanical components are fabricated using compatible micromachining. Microelectronic-integrated circuits can be thought of as the "brains" of a system.

In order to solve the measurement problem in the harsh environment, a sensor has been developed, which can be used under high temperature, high pressure, and is able to endure instantaneous ultra high environmental impact. Based on the MEMS and IC technology, the sensor's sensitive element was fabricated and constituted by silicon substrate. The experimental results show that the developed sensor has good performances under critical circumstances and is able to endure instantaneous ultra high temperature/pressure/flow impact, which meets the requirements of modern industry, such as aviation, oil, engine, etc.

Sensors made using MEMS are better than their conventional counterparts because they are:

- Smaller in size
- Have lower power consumption
- More sensitive to input variations
- Cheaper due to mass production
- Less invasive than larger devices.

The sensor is fabricated with a series of standard silicon processing techniques. *Silicon* is the standard substrate material for IC fabrication and, hence, the most common substrate material in fabrication in general. A wafer is a thin slice of a silicon crystal (semiconductor material), used in the fabrication of IC (i.e., sensor) and other micro devices. The wafer serves as the substrate for microelectronic devices built in and over the wafer and undergoes many micro fabrication process steps such as doping or ion implantation, etching, deposition of various materials, and photolithographic patterning [14].

MEMS applications:

- Accelerometers,
- silicon-based Piezo-resistive pressure sensors,
- capacitive pressure sensors,
- digital micro mirror arrays,
- chemical/biosensors and gas sensors,
- micro resonators,
- infrared detectors,
- focal plane arrays for earth observations,
- space science,
- missile defense applications,

- Pico-satellites for space applications,
- fuel cells, and many hydraulic, pneumatic, and
- Other consumer products.

Fabrication process:
The various process steps needed to be carried out for realizing the sensor using bulk/surface micromachining approach. The wafer serves as the substrate for microelectronic devices built in and over the wafer and undergoes many micro fabrication process steps such as doping or ion implantation, etching, deposition of various materials, and photolithographic patterning.

Oxidation:
The process of oxidation consists of growing a thin film of silicon dioxide on the surface of the silicon wafer. Oxidation is typically performed at temperatures of 900–1,200 °C in the presence of O_2 (dry oxidation) or H_2O (wet oxidation).

Etching Techniques:
Etching is the process of selective removal of regions of a semiconductor, metal, or silicon dioxide. There are two types of etchings: wet and dry/Isotropic versus Anisotropic. Figure 2.20 displays the schematic of isotropic and anisotropic thin-film etching.

In wet etching, the wafers are immersed in a chemical solution at a predetermined temperature. In this process, the material to be etched is removed equally in all directions so that some material is etched from regions where it is to be left. This becomes a serious problem when dealing with small dimensions.

In dry (or plasma) etching, the wafers are immersed in gaseous plasma created by a radio-frequency electric field applied to a gas such as argon. Electrons are initially released by field emission from an electrode. These electrodes gain kinetic energy from the field, collide with, and transfer energy to the gas molecules, which results in generating ions and electrons. The newly generated electrons collide with other gas molecules and the avalanche process continues throughout the gas, forming plasma. The wafer to be etched is placed on an electrode and is subjected to the bombardment to fits surface by gas ions. As a result, atoms at or near the surface to be etched are removed by the transfer of momentum from the ions to the atoms.

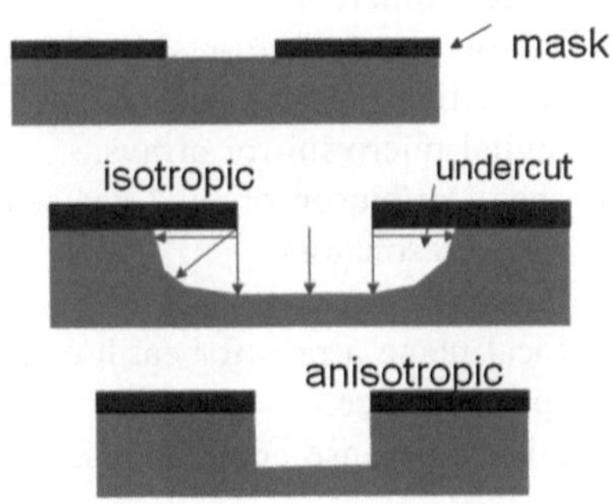

Fig. 2.20 Schematic of isotropic and anisotropic thin-film etching

Doping/Diffusion:
The introduction of certain impurities into a semiconductor can change its electrical, chemical, and even mechanical properties. This process consists of the introduction of a few tenths to several micrometers of impurities by the solid-state diffusion of dopants into selected regions of a wafer to form junctions. Most of these diffusion processes occur in two steps: the *pre-deposition* and the *drive-in* diffusion. In the pre-deposition step, a high concentration of dopant atoms are introduced at the silicon surface by a vapor that contains the dopant at a temperature of about 1,000 °C. Diffusion is an isotropic process, the doped area will also extend underneath the mask. In micro-fabrication, diffusion is mainly used in the formation of very highly doped boron regions (p++), which are usually used as etch steps in bulk micromachining. In recent years Ion Implantation is used.

Ion Implantation:
This is a process of introducing dopants into selected areas of the surface of the wafer by bombarding the surface with high-energy ions of the particular dopant. In ion implantation, the impurities are ionized and accelerated toward the semiconductor surface.

After implantation, an annealing process is needed to activate the impurities and to repair the damage in the crystal structure generated by the ion collisions. A *drive-in* process to redistribute the impurities, performed in a standard furnace like those used for oxidation or diffusion may also be required.

Ion implantation allows more precise control of the dose (total amount of impurities introduced per area unit) and the impurity profile (concentration versus depth).

Photolithography:
Photolithography is the standard process to transfer a pattern, which has been designed with computer-aided-engineering (CAE) software packages, on to a certain material. Basically, photolithography comes down to applying a layer of a light-sensitive material (photoresist) on a flat substrate, illuminating it by some source using some pattern (a mask) thereby making the illuminated parts either soluble or insoluble and then removing the soluble parts [15].

Epitaxy:
Epitaxy is the process of the *controlled growth* of a crystalline doped layer of silicon on a single crystal substrate. Epitaxy is used to deposit Non N+ silicon, which is impossible to accomplish by diffusion. It is also used in isolation between bipolar transistors where in N− is deposited on P.

Selective epitaxial growth is of particular interest for the formation of microstructures.

Metallization and interconnections:
After all semiconductor fabrication steps of a device or of an IC are completed, it becomes necessary to provide metallic interconnections for the IC and for external connections to both the device. Figure 2.21 shows the example of fabrication process implemented in accelerometer.

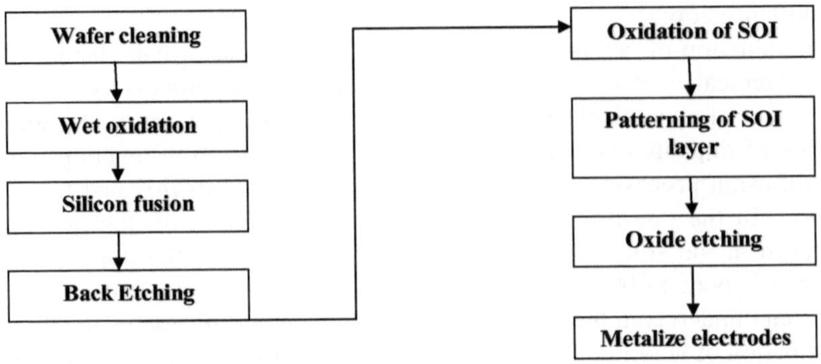

Fig. 2.21 Example: fabrication process of accelerometer

References

1. W.M. Richard, A sensor classification scheme. IEEE Trans. Ultrason. Ferroelectr. Freq. Control, **UFFC-34**(2) (March 1987)
2. J. David Zook, N. Schroeder, in *Sensors as Information Transducers*. Published in: Encyclopedia of Sensors, ed. by C.A. Grimes, E.C. Dickey, M.V. Pishko vol. 9 (American Scientific Publishers, Stevenson Ranch, California) pp. 329–359
3. NSDL OAI Repository, http://www.analog.com/library/analogDialogue/archives/42-03/CH03-H8703.pdf
4. WIKA Handbook, Pressure and Temperature Measurement, www.mika.com
5. NSDL OAI Repository, http://www.omega.com/temperature/Z/pdf/z021-032.pdf
6. Basic instrumentation measuring devices and basic PID controls.pdf, Science and Reactor Fundamentals. Instrumentation & Control CNSC Technical Training Group
7. NPTEL lecture series, Version 2, IIT, Kharagpur. http://nptel.iitk.ac.in/
8. F.S. Tse, I. Morse, *Measurement and Instrumentation in Engineering* (Marcel Dekker, New York, 1989)
9. Technical Guide Book: Flow Sensors, www.keyence.com
10. NSDL OAI Repository, http://www.capsense.com/capsense-wp.pdf
11. Guide to Displacement Sensor, www.keyence.com
12. Z. Chen, L. Chi, Humidity sensors: a review of materials and mechanisms. Sens. Lett. **3**, 274–295 (2005)
13. Instruments Engineers' Handbook, National nuclear security administration (nevada). http://www.jcte.jcs.mil/
14. G.K. Fedder, MEMS Fabrication. Department of Electrical and Computer Engineering, and The Robotics Institute Carnegie Mellon University, Pittsburgh, USA
15. http://www.utwente.nl/ewi/tst/education/el-bach/mandt/extra/background/mems_sensors.pdf
16. Technical Guide: Pressure Sensor, www.keyence.com
17. Various techniques of liquid and solids level measurements, Indumart Newsletter (2009)

Chapter 3
Biological Olfaction

3.1 Introduction

Olfaction is an often over-looked but vital sense to our everyday living. Olfactory information can influence behavior, social interactions, and in many animals, reproduction. Humans, who rely on smell less than many other mammals, can distinguish between 5,000 and 10,000 different odorants. Olfaction has an extremely high importance in the human being. It is one of the five main senses: Sight, Smell, Taste, Hearing and Touch [1]. There are multiple occupations for which olfaction is depended on for making an accessible. These include chefs, firefighters, plumbers, wine merchants, perfumers, cosmetic retailers, and chemical plant workers.

3.2 The Nose: How It Works

The human nose is in fact the main organ of taste as well as smell. The so-called taste-buds on our tongues can only distinguish four qualities—sweet, sour, bitter and salt -all other "tastes" are detected by the olfactory receptors high up in our nasal passages.

Anything that has an odor constantly evaporates tiny quantities of molecules that produce the smell, so-called odorants. A sensor that is capable to detect these molecules is called a *chemical sensor*. In this way the human nose is a chemical sensor and the smell is a chemical sense [2]. The nose is an important regulator of social life. Pheromones are a class of long distance chemical messenger hormones that regulate social relations, behavioral and physiological responses in insects and many mammals.

The mammalian olfactory system regulates a wide range of multiple and integrative functions such as physiological regulation, emotional responses (e.g., anxiety, fear, pleasure), reproductive functions (e.g., sexual and maternal behaviors), and social behaviors (e.g., recognition of cheat specifics, family, clan, or

H. K. Patel, *The Electronic Nose: Artificial Olfaction Technology*,
Biological and Medical Physics, Biomedical Engineering,
DOI: 10.1007/978-81-322-1548-6_3, © Springer India 2014

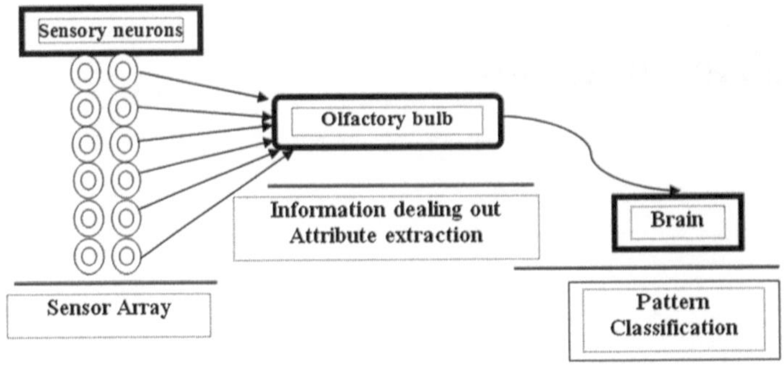

Fig. 3.1 Functional components of olfactory system

outsiders). The architecture and dynamics of the olfactory system have evolved to solve the major problems in olfactory information processing.

The chemical sensors in the olfactory system act as receptors and the response forms a pattern (a set of signals) that is processed in the brain to produce an appropriate response due to a stimulus. Figure 3.1 shows a block diagram of the olfactory system. In a human, there is an array of 100 million odor sensors. Within this array, there are a number of different types of sensors, which display differing odor specificity to particular classes of odors. Individual elements in the array show broad and overlapping selectivity to chemical species. Importantly, olfactory receptors are not highly selective, but selectivity is achieved by the unique patterns of responses from numbers of such receptors.

Taste and smell are highly associated with each other, and flavors are composite sensations, derived from primary taste and smell sensations by processes within the limbic brain areas that participate in emotion. The human emotional system emerged in evolutionary terms from the olfactory brain of earlier animals and remains closely linked to taste, smell, and eating behaviors. Taste and smell evaluation may be an important diagnostic screen for early brain changes destined to result in serious dementia [3]. The odors of coffee, chocolate, almond and oil of lemon are often used for casual testing of olfactory function. Diversity of different methods has been used to understand olfaction. Operations of olfaction can be divided roughly in five parts: sniffing, reception, detection, recognition, and cleansing.

The olfaction starts with *sniffing* that mixes the odorants into a uniform concentration and delivers these mixtures to the mucus layer in the upper part of nasal cavity. Next these molecules are dissolved in this layer and transported to the cilia of the olfactory receptor neurons. *Reception* process includes binding of these odorant molecules to the olfactory receptors. Odorant molecules are binded temporarily to proteins that transport molecules across the receptor membrane with simultaneous stimulation of the receptors [4]. During this stimulation the chemical reaction produces an electrical stimulus. These electrical signals from the receptor neurons are transported to the olfactory bulb. From the olfactory bulb the receptor response information is forwarded to the olfactory cortex (*detection*). Odor *recognition* part takes place namely in the olfactory cortex. Then the information is transmitted to the cerebral cortex. Remind that there are no individual receptors or parts of the brain capable to recognize specific odors. The brain is key component for the collection of olfactory signals associated with the specific odor [4]. *Cleansing* finishes the olfaction process. For this purpose the breathing fresh air removing of odorant molecules from the olfactory receptors is required. To grasp the mechanism of olfactory perception the model of our nose can be considered.

3.3 Human Olfaction System

The knowledge of the mammalian olfactory system is deficient, but researchers are making enormous progress by conducting experiments in molecular biology. Odors are our perceptions of mixtures of volatile organic compounds (VOC). These mixtures are carried in by inhaled air to the olfactory region in the nasal cavities. Information that is used constantly in daily life—about food, dangerous substances, and potential mates are provided by smell. VOCs, which are either produced endogenously during metabolism are inhaled with the breathed ambient air and consequently modified during a toxic kinetic process in the body. Figure 3.2 show correlations between particular diseases, albeit accompanied in most cases [16].

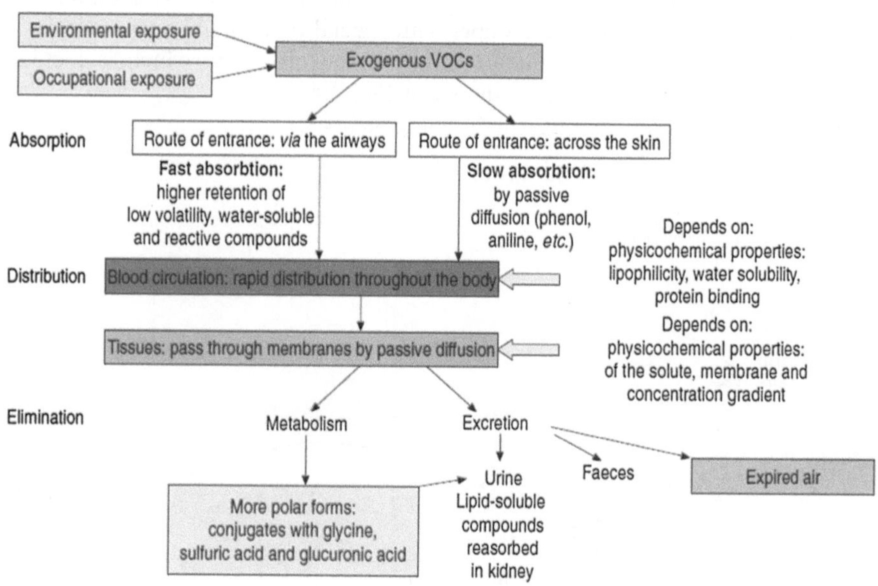

Fig. 3.2 Toxicokinetics of exogenous volatile organic compounds (VOCs) in the human body [16]

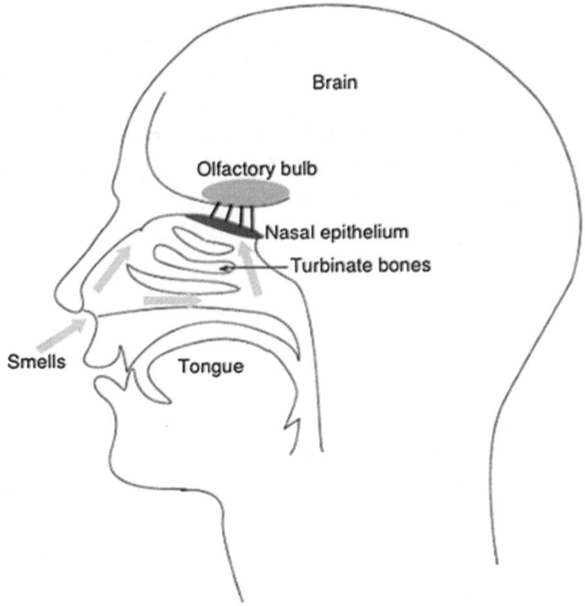

Fig. 3.3 Olfactory system in human

The olfactory sensors are located in the olfactory epithelium in the roof of the two nasal cavities between the eyes as shown in Fig. 3.3. Odor sensations are produced by the interaction of the odors with special trans-membrane receptors in the olfactory epithelium. Schematic of the human olfactory, which is includes olfactory receptors, olfactory bulb shown in Fig. 3.4.

Olfactory system has its three main components.

1. The layer of receptors in the nose.
2. The olfactory bulb.
3. The prepyriform cortex.

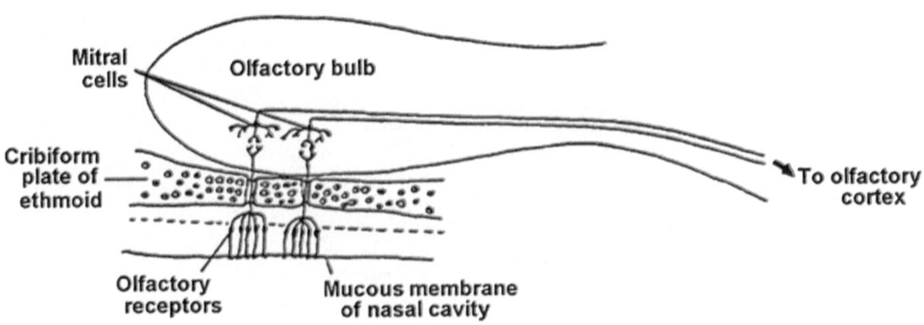

Fig. 3.4 Schematic of the olfactory receptors, olfactory bulb, olfactory tract

3.3.1 Olfactory Receptors

Receptor cell as shown in Fig. 3.5, embedded in a specialized patch of yellow-tinted mucous membrane in the roof of the nasal cavity is called olfactory receptors. It's only neurons that regularly die as its life span is 5–8 weeks and are replaced by new one. There are 10–20 million receptor cells spread in the middle of the supporting cells of the mucous membrane. Dendrites defined as extensions of receptor cells have to carry electrical signals into the cell body. It has cilia that is projecting from the dendritic knob into the surface of the mucus layer which has micro tubular hair like structures and on which the receptors for odorants are located. Cilia contain the active sites for the olfactory transduction process. Figure 3.6 shows the basic neuron. A neuron which is bipolar and covered with non-motile cilia called receptors. Axons from the olfactory receptors enter small nerve bundles (collectively termed

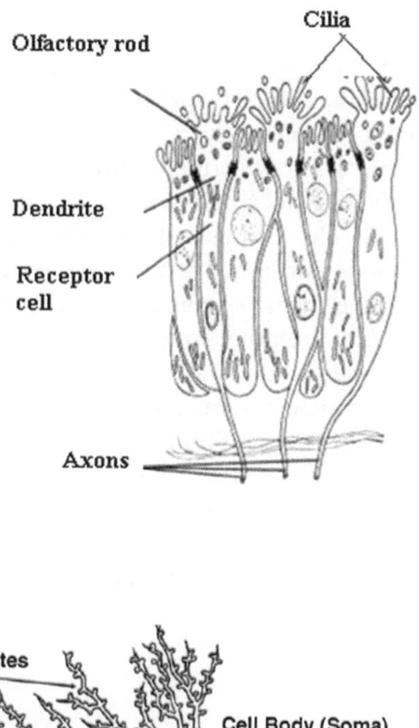

Fig. 3.5 Receptor cell

Fig. 3.6 A neuron

the 1st cranial nerve) which pass through the perforations in the cribi form plate of the ethmoid bone and promptly enter the olfactory bulb. These nerve bundles can be severed as a result of skull fractures or other pathology in this region with a resulting partial or complete anosmia (loss of sense of smell). Much of the sensation which is considered to be taste is actually olfactory so patients with anosmia often complain bitterly about loss of pleasure from eating.

Certain types of chemicals stimulate the receptor cells of smell and taste. The major functional difference between the two kinds of receptors is that taste receptors are specialized cells that detect chemicals present in quantity in the mouth itself, while smell receptors are modified sensory neurons in the nasal passages that detect vapors coming from distant sources. The smell receptors can be as much as 3,400 times more sensitive than the taste receptors.

3.3.2 Olfactory Bulb

The olfactory bulbs lie on the ventral aspect of the frontal lobes. The olfactory bulbs and all other parts of the olfactory pathways are telencephalic derivatives. Within the olfactory bulbs the olfactory nerves synapse on mitral cells whose axons project directly to the olfactory cortex. *Mitral cells* are the principal neurons in the olfactory bulb. In the olfactory bulbs, the axons of the receptors terminate and form the olfactory glomeruli. Averages of 26,000 receptor cell axons converge on each glomerulus. Impulses concerned with olfactory reflexes are passed from the glomeruli to the rest of the limbic system and the hypothalamus. The olfactory bulb carries out a great deal of pre-processing and feature extraction of the signals arriving from the olfactory receptors, analyzing each input pattern and then producing specific messages, which it transmits via axons to another part of the olfactory system, the olfactory cortex. From there, new signals are sent too many parts of the brain, including an area called the entorhinal cortex, where the signals are combined with those from other sensory systems. The result is a meaning-laden perception which is unique to each person, where some scents may produce a sense of well-being, and others a sense of nausea, and others may be linked to specific memories or events (Fig. 3.7).

3.3.3 Olfactory Cortex

Those portions of the cerebral cortex that receive direct projections from the olfactory bulb (via mitral cell axons) are collectively referred to as the olfactory cortex. Note that the olfactory cortex is the one area of cortex that receives direct sensory input without inter posed thalamic connection. Most of the olfactory cortex is of a primitive 3-layered type. The olfactory cortex is located on the base of the frontal lobe and medial aspect of the temporal lobe (Figs. 3.8, 3.9).

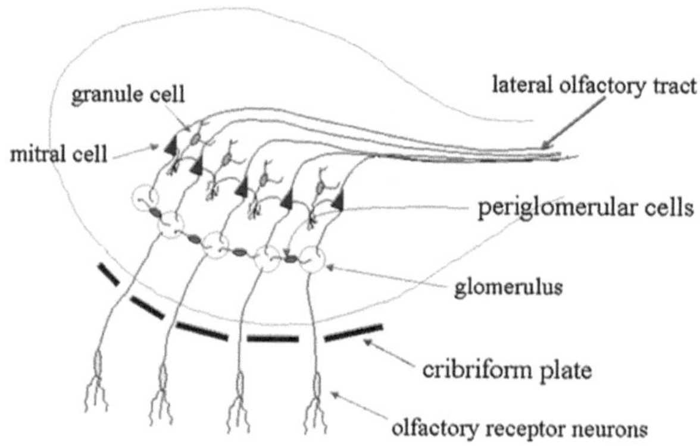

Fig. 3.7 Olfactory bulb [17]

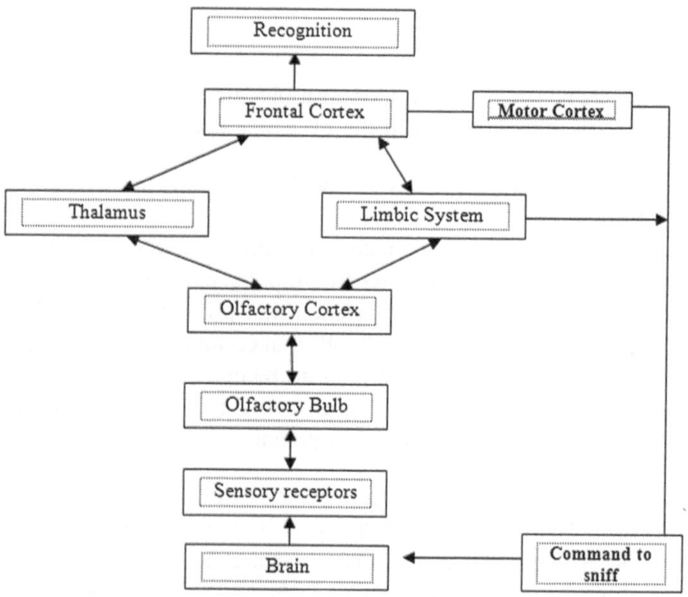

Fig. 3.8 Olfactory pathways

3.3.4 Organization of Human Olfactory System

The olfactory cortex is located on the base of the frontal lobe and medial aspect of the temporal Lobe. On the base of the frontal lobe it overlies the anterior perforated substance through which the striate arteries enter the interior of the brain. From the olfactory cortex, olfactory information is relayed via the mediodorsal nucleus of the thalamus to the insular and orbitofrontal cortex.

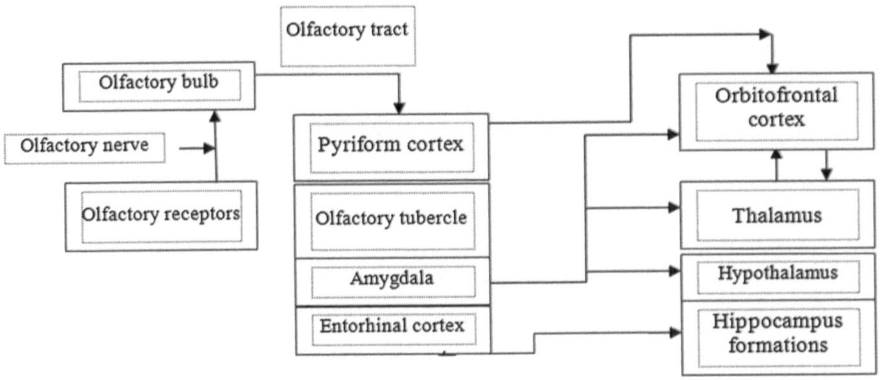

Fig. 3.9 Odorants pathway

3.3.5 Olfactory Epithelium

The sense of smell begins with an odorant molecule binding to a receptor on the cilia of olfactory receptor neurons in the olfactory epithelium.

1. The olfactory epithelium resides in the back of the nasal cavity below the olfactory bulbs.
2. Roughly 5–10 million olfactory receptor neurons (ORNs) reside in the olfactory epithelium. These receptors are true bipolar neurons, extending a dendrite toward the epithelial surface and an axon that project into the olfactory bulb. The ends of the ORN dendrites are knobs that contain fine cilia projecting into a layer of mucus. On these cilia the odorant receptors are located. Odorants are dissolved in the mucus layer and bind to the receptors, initiating the transduction process that leads to an action potential in the ORN.

Humans and other mammals are capable of discriminating a great variety of odors and flavors. The olfactory capability of humans is somewhat limited compared with that of some other mammals; we are nevertheless able to perceive thousands of different odorous molecules (*odorants*). Perfumers, who are highly trained to discriminate odorants, say that they can distinguish as many as 5,000 different types of odorants, and wine tasters report that they can distinguish more than 100 different components of taste based on combinations of flavor and aroma.

3.4 Olfactory Transduction

The information about conditions inside or outside the body is provided to the central nervous system with the cell processes (i.e., Sensory receptors). Our sensitivity to temperature, pain, touch, pressure, vibration is described as generally senses. Sensory receptors are distributed throughout the body. Sensory information is interpreted on the basis of the frequency of arriving action potentials. Every receptor has a distinctive sensitivity. i.e., a touch receptor is sensitive to pressure but comparatively insensitive to chemical stimuli, while a taste receptor is sensitive to dissolved chemicals but insensitive to pressure. Various classifications of human sensory system and its associated stimuli are shown in Table 3.1.

The purpose of transduction is to convert a pattern of information distributed over space and time to a pattern of neural activity which is also appropriately distributed across space (different neurons) and time. The process through which a specific pattern of information (energy) in the environment (e.g., light, vibrations, dissolved chemicals or airborne chemicals) is converted to a pattern of electrical activity in the nervous system is called *transduction*. (Conversion of a chemical signal (odor binding) to an electrical signal [change in neuron's V_m)].

In human olfactory process when odorant molecules reach the olfactory mucosa and bind to the olfactory receptor proteins on the cilia of the olfactory receptor neurons olfactory transduction occurs. Each olfactory receptor neuron has multiple types of receptor sites for different molecules [5]. The various types of receptor sites are distributed in different proportions on different olfactory receptor neurons. Binding of odorants to receptor sites associated with ion channels causes a receptor potential at the dendrites, and action potentials at the cell body and axon. The axons of olfactory receptor cells project directly to the brain via the olfactory nerve.

Table 3.1 Classification of human sensory system

Type of sensory system	Modality	Adequate stimuli
Mechanical	Touch	Contact with body surface
	Hearing	Sound vibrations in air or water
	Vestibular	Head movement and orientation
	Joint	Position and movement
	Muscle	Tension
Photic	Seeing	Visible radiant energy
Thermal	Cold	Skin temperature decrease
	Warmth	Increase in skin temperature
Chemical	Smell	Odorous substances dissolved in air or water in the nasal cavity
	Taste	Substances in contact with the tongue or other taste receptor
	Common chemical vomeronasal	Changes in CO_2, pH, osmotic pressure
Electrical	Electroreception	Differences in density of electrical currents

3.4.1 Steps in Olfactory Transduction

Odorant molecules → Golf → adenylate cyclase → cAMP → protein Kinase → Ion Channel Protein → Membrane Conductance change → Receptor potential → Electronic propagation → Action potential.

When odorants bind to the receptor site, the receptor protein changes shape which in turn triggers the flow of ions across the receptor-cell membrane and an electrical response is triggered in the cilium. Electrical responses in the cilia spread to the rest of the receptor cell, and from there are passed onto the olfactory bulb of the brain in the olfactory nerve. There are about 1,000 different types of receptor proteins each sensitive to different odorants. Human have a total of about 10 million receptor neurons. Each receptor neuron has about 1,000 similar receptor proteins. Because there are 1,000 different receptor proteins, there are also 1,000 different receptor neurons.

The olfactory systems gather at the interface of the environment and the nervous system. The olfactory system is responsible for correctly coding sensory information from thousands of odorous stimuli. Many theories existed regarding the signal transduction mechanism that mediates this difficult task. The discovery that odorant transduction utilizes a unique variation (a novel family of G protein–coupled receptors) based upon a very common theme of the G protein–coupled adenylyl cyclase cascade to accomplish its vital task emphasized the power and versatility of this design. Within the compact cilia of the OSNs a cascade of enzymatic activity transduce the binding of an odorant molecule to a receptor into an electrical signal that can be transmitted to the brain. Sensory transduction is shown in Fig. 3.10. As described in detail, this is a classic cyclic nucleotide transduction pathway in which all of the proteins involved have been identified, cloned, expressed and characterized.

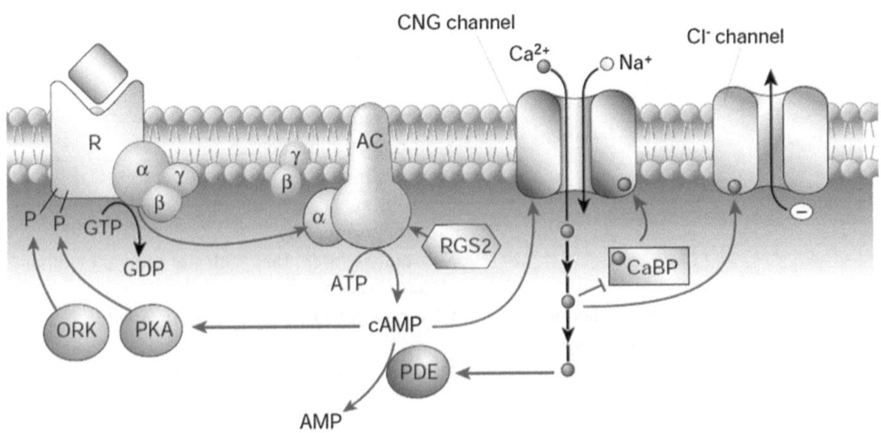

Fig. 3.10 Sensory transduction [12]

AC—adenylyl cyclase; CNG channel—cyclic nucleotide-gated channel; PDE—phosphodiesterase; PKA—protein kinase A; ORK—olfactory receptor kinase; RGS—regulator of G proteins (but here acts on the AC); CaBP—calmodulin-binding protein. Green arrows indicate stimulatory pathways; red indicates inhibitory (feedback).

The odor molecule has bound with receptor; a cascade of events is initiated that transforms the chemical energy of binding into a neural signal that is due to change in the membrane potential of the olfactory sensor neuron. Even though still difficult to understand this process but it is mainly well understood in mammals and humans. G-s family of protein (Family of G proteins) Receptor activates by receptor which in turn activates adenylyl cyclase. The abundant intracellular molecule ATP is converted into cyclic AMP (molecule that has numerous signaling roles in cells) by the cyclase. The cyclic AMP binds to the intracellular cation selective (Na^+, K^+, Ca^{2+}) channel from the inside of the membrane. This channel is known as cyclic-nucleotide gated (CNG) channels which is closely related and found in photoreceptors. The influx of Na^+ and Ca^{2+} ions causes the inside of the cell to become less negative when the CNG channels are open. If required CNG channels are open for long time, causing the membrane potential to become less negative and near about the cell threshold and generates an action potential. The axon circulated the action potential which crosses through a thin bone known as the cribiform plate, and into the forebrain where it synapses with second order neurons in the olfactory bulb. Cascade of enzymes provides amplification and integration of odor-binding events done by second messenger. A bound odor activated one membrane receptor which in turn activates tens of G proteins, each of which will activate a cyclase molecule capable of producing about a thousand molecules of cAMP per second.

To open a channel, cAMP molecules are required but hundreds of thousands of ions can cross the membrane through a single open channel. So, to produce a measurable electrical event in an OSN few channels opening together required and it could pass sufficient current to induce action potential generation [6, 7]. Addition to this pathway, a somewhat unique amplification mechanism is implemented in OSNs. Through CNG channel the calcium ions are entering and it's able to activate another ion channel that is absorbent to the negatively charged chloride ion [8].

Inhibitory responses mediate in Neuronal through Cl^- channels, and Cl^- ions enter the cell through an open channel. OSNs maintain high intracellular Cl^- concentration such that there is a Cl^- efflux when these channels are activated. Net positive charge is on the left behind of membrane that further depolarizes the cells, so adding to the excitatory response magnitude. This interesting revision to the fact that the olfactory cilia reside in the mucus, outside the body proper and where concentrations of ions are not as well regulated as they are in normal interstitial compartments [9, 10]. Thus the OSN maintain its own Cl^- battery, if in case the insufficient threshold current is not provided by Na^+ gradient in the mucus, this will uses to boost the response.

Through a negative feedback pathway involving Calcium ions entering through the CNG channels are also important in response adaptation [11]. A calcium ion acts on the channel during the odor response and decrease its sensitivity to cAMP, so requiring odor stimulus to produce sufficient cAMP to open the channel. As in vertebrates, invertebrates have both excitatory and inhibitory responses to odors, so there are likely to be multiple transduction pathways.

3.5 Encoding of Smell

The ability to measure things is a critical component in scientific inquiry. If you're studying the strength of scents and their impact on people, you need to come up with a standardized unit of what exactly constitutes the sensory input of smell. To that end, Danish professor Dr. P. Ole Fanger created the smell measurement unit known as an "Olf", short for "olfaction unit" derived from the Latin word "olfactus" which means smelled. 1 Olf is the strength of the smell produced by a healthy adult working a sedentary job, in a climate controlled environment, with the hygienic equivalent of 0.7 baths per day, and a skin surface area of 1.8 m^2.

Using that as a base measure, the "Olf" factor of other things can be determined. A heavy smoker, for example, generates 25 Olfs. An athlete after participation in a strenuous sport generates 30 Olfs. A non-porous and virtually scentless material like polished Marble gives off a mere 0.01 Olfs per square meter of surface area.

The olfactory receptor nerves located at the upper limit of the nasal cavity out of the airflow channel. Together with mucous-secreting cells and basal cells, they make up the odor detection apparatus. The receptor portions of the olfactory neuron resemble cilia that extend to the mucosa surface. The finding sites on these ciliated nerves await specific molecules that diffuse from the inspired airstream under normal breathing. Direct transport of air is affected by sniffing that creates turbulence and opens up passages by the turbinate's. Switching to oral breathing can mitigate the intensity of malodorous compounds. When reaching the mucosa, odorous molecules must be dissolved before binding to activate a neural impulse. The impulse moves into the olfactory bulb in the brain, which is a complex neural structure. This second-stage processed signal then travels to the limbic system and the thalamus- cortex regions of the brain. The limbic cortex region affects behavioral reactions associated with smell, whereas the thalamus-cortical region is conscious interpretation of smell. Many of us share the experience of recalling a memory when we encounter a particular odor.

Odor encoding is a spatially distributed process. Single odors can activate broad overlapping regions, it is clear that individual neurons—and thus possibly the receptor proteins themselves—can be activated by many odorants, including ones that belong to different chemical families, underlying different odor qualities. Because odorants are such complicated stimuli, we don't yet have a complete picture of how smell is encoded by the brain. Different areas of the mucosa are sensitive to different types of odorants. Smells appear to be organized spatially in the olfactory bulb (similar smells are grouped together). An odotope is a group of odorants that share some chemical feature and cause similar patterns of neural firing. Neurons that fire to the same odotope are usually located near each other.

It is hard to imagine how to build a system that could reproduce smell information collected by human olfactory system. This is sometimes discussed in terms of a multi-dimensional "odour space," where each odor could be described by its location along each dimension. It is clear that odors are comprised of chemicals, but still it's under research. Understanding the odor space is a prerequisite to build a system that can transmit messages from it.

Also, to a great extent, how to engineer a smell transmission system is only partly a biological question. Transmitted pictures and transmitted sounds can be coded because they activate our photoreceptors and auditory hair cells in the same pattern that the original images and sounds would. We can build systems for picture and sound because we understand the nature of the information sensory systems decode, and we don't need a really deep understanding of the physiology to do that.

3.6 Signal Transduction System

The cell senses extra cellular signals: –Hormones, pheromones, heat, cold, light, osmotic pressure, concentration change of glucose, K^+, Ca^{2+}or cAMP and compute them in intercellular signals.

One of the main functions of neurons is to communicate with other neurons. An individual neuron may receive information from many different sources. Its job is to evaluate this information and "make a decision" as to whether to send out information to all of its target neurons, or whether to remain silent. It is hard to know where to begin describing the process of signaling in neural circuits, because signaling in any individual neuron depends on its getting input from other neurons that have been activated by the same processes that will consider in model neuron.

Signal Transduction—chain of events that converts the message "this molecule is present" to a physiological response.

Key Steps in Signal Transduction:

- Release of the primary messenger
- Reception of the primary messenger
- Delivery of the message inside the cell by the second messenger

Electric signal in nerve: nerve cells or neurons make up the nervous system (NS), one of the control systems of the body. The nervous system controls body's muscular and glandular activities that are mostly directed towards maintaining homeostasis. Neurons act rapidly for electrical and chemical signaling for communication. Through chemical means neurons pass messages to muscles and glands through intricate pathways from neuron to neuron. Nerve and muscle are excitable tissues. They are able to develop rapid and transient change in their membrane potentials.

The process of neural signaling is, in a sense, a cyclic process, so it can be described anywhere in the cycle and eventually come around to the starting point again. One logical point at which to start is the process of *synaptic transmission*, the transfer of information from one neuron (the *presynaptic* neuron) to another (the *postsynaptic* neuron) by release of a chemical substance, or *neurotransmitter*, from the *axon terminal(s)* of the presynaptic neuron. The neurotransmitter affects the state of the postsynaptic neuron, making it more or less likely to transmit information to its targets.

Before any neural processing can take place, there must be some means of detecting the presence of information in the environment, collecting the different forms and patterns of energy that represent this information, and converting the physical energy into a form that can be acted upon and utilized by the nervous system. The process through which a specific pattern of information (energy) in the environment (e.g., light, vibrations, dissolved chemicals or airborne chemicals) is converted to a pattern of electrical activity in the nervous system is called *transduction*.

Signal transduction occurs in response to a signal and consists of a series of post-translational modifications that regulate the activity of proteins. The net result of this regulatory cascade is a change in cell activity. This change in cell activity might involve the production of a product, such as activation of the transcription of a gene. But it might not. It might just result in a different cytoskeletal configuration, vesicle fusion or could result in the repression of a gene's transcription. Lots of other responses are possible, including cell division, differentiation, or death. So the easiest way to think about signal transduction pathways is simply as a mechanism (cascade of posttranslational modifications) that regulates a cellular activity or activities in response to a signal.

Olfaction (smell), vision (sight), gustation (taste), equilibrium (balance), hearing, somato-sensory (touch, pain, temperature) are the senses of human system. Special sensory receptors are located in sense organs such as the eye or ear, where the receptors are protected by surrounding tissues. The information these receptors provide is distributed to specific areas of the cerebral cortex (the auditory cortex, the visual cortex, and so forth) and to centers throughout the brain stem. Table 3.2 enlists the sensory system receptor and transduction energy used in the sensory system.

Beginning with the physical energy of an environmental stimulus, there are several processes that must occur before a pattern of neural activity is generated. The steps leading to transduction include the following:

The stimulus energy must reach specialized receptor cells. In taste & touch this process is relatively simple and straightforward.

Table 3.2 Human sensory system

Sensory system	Modality	Stimulus energy	Receptor class
Visual	Vision	Light	Photoreceptor
Auditory	Hearing	Sound	Mechanoreceptor
Vestibular	Balance	Gravity	Mechanoreceptor
Somato sensory	Somatic sensors:		
	Touch	Pressure	Mechanoreceptor
	Proprioception	Displacement	Mechanoreceptor
	Temperature sense	Thermal	Thermoreceptor
	Pain	Chemical, thermal or mechanical	Chemoreceptor Thermoreceptor
Gustatory	Taste	Chemical	Chemoreceptor
Olfactory	Smell	Chemical	Chemoreceptor

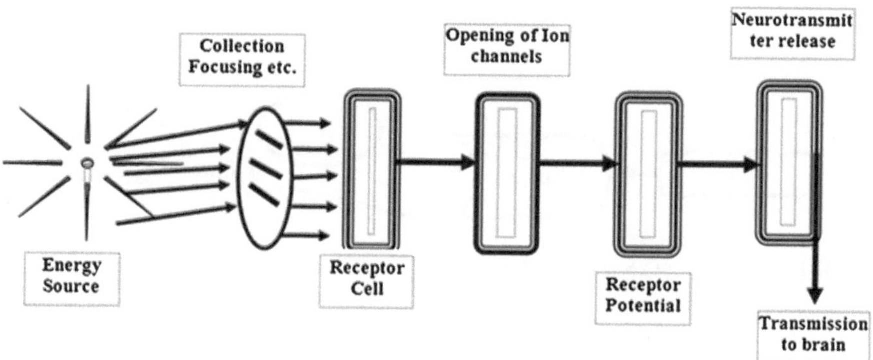

Fig. 3.11 Diagram summarizing the events that take place in transduction

The receptor cells must be activated. The activation process involves opening of ion channels to cause a change in the cell's membrane potential. In different sensory systems, different mechanisms cause ion channels to open in response to a stimulus.

Figure 3.11 summarizes the events that take place during the transduction in sensory system. The opening of ion channels creates a *receptor potential*. Like an EPSP in a neuron, the receptor potential is graded; its size reflects in some way the properties of the stimulus. In general, larger the magnitude of the stimulus, larger is the receptor potential. The receptor potential causes release of neurotransmitter onto the dendrite of the *primary afferent* (nerve fiber projecting to the central nervous system). The larger the receptor potential, the greater the quantity of neurotransmitter released. If enough excitatory neurotransmitter is released to bring the primary afferent neuron to threshold, it will fire an action potential.

3.6.1 Characteristics of Signal Transduction Events

Specificity: Signal molecule fits binding site on its complementary receptor; while other signals do not fit. Figure 3.12 illustrates the receptor binding of signal molecules.

Amplification: when enzymes activate enzymes, the number of affected molecules increases geometrically in an enzyme cascade. Figure 3.13 illustrate amplification process of enzymes.

Adaptation: Receptor activation triggers a feedback circuit that shuts off the receptors or removes it from the cell surface. This process is illustrated in Fig. 3.14.

Integration: when two signal have opposite effects on a metabolic characteristic such as the concentration of a second messenger X, or the membrane potential V_m, the regulatory outcome results from the integrated input from both receptors. Signal integration is shown in Fig. 3.15.

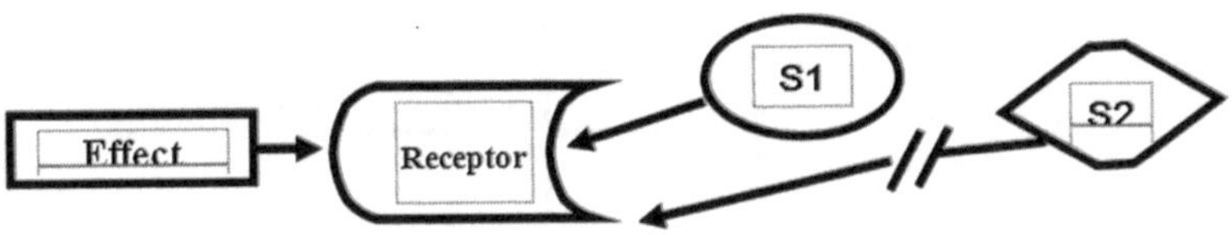

Fig. 3.12 Illustration of receptor binding

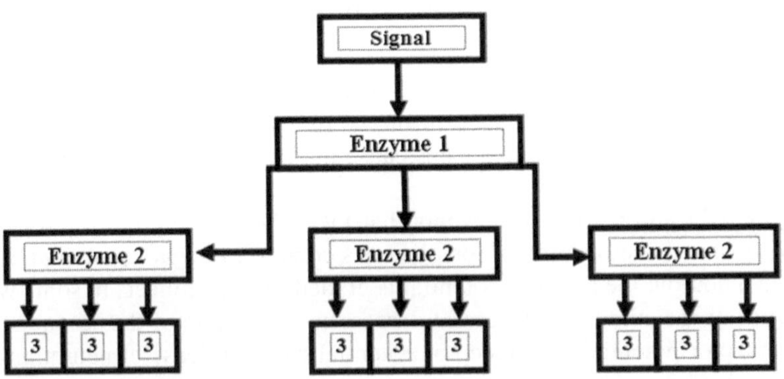

Fig. 3.13 Enzyme amplification

Fig. 3.14 Adaptation

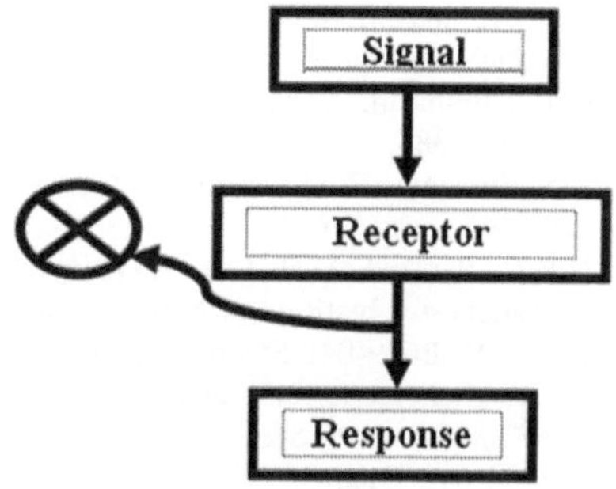

Fig. 3.15 Signal transduction → external signal → receptor → transducer amplifier→ phophorylated precursor → 2nd messenger → internal effectors membrane response cellular response

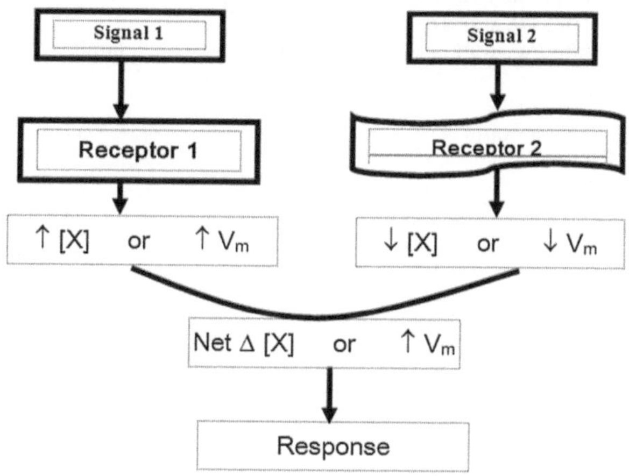

References

1. The Senses. http://users.tpg.com.au/users/amcgann/body/senses.html. (Cited 29 Oct 2003)
2. A. Mamlouk, *Quantifying Olfactory Perception*, Master of Science Thesis, University of Lubeck, Germany, 2002
3. F. Wortmann, Receptor multi-specificity and similarities between the immune system, the sense of smell, and the nervous system from the point of view of clinical allergology. Ann. Allergy **59**, 65–73 (1987)
4. P. Keller, Overview of electronic nose algorithms. International joint conference of neural networks (IJCNN'99), Washington, USA, 1999
5. J.P. Rospars et al., Odour transduction in olfactory receptor neurons. Chin. J. Physiol. **53**(6), 364–372 (2010). doi:10.4077/CJP.2010.AMM038
6. J.W. Lynch, P.H. Barry, Action potentials initiated by single channels opening in a small neuron (rat olfactory receptor). Biophys. J. **55**, 755–768 (1989)
7. A. Menini, C. Picco, S. Firestein, Quantal-like current fluctuations induced by odourants in olfactory receptor cells. Nature **373**, 435–437 (1995)
8. S.J. Kleene, R.C. Gesteland, Calcium-activated chloride conductance in frog olfactory cilia. J. Neurosci. **11**, 3624–3629 (1991)
9. G. Lowe, G.H. Gold, Nonlinear amplification by calcium-dependent chloride channels in olfactory receptor cells. Nature **366**, 283–286 (1993)

10. T. Kurahashi, K.W. Yau, Co-existence of cationic and chloride components in odourant-induced current of vertebrate olfactory receptor cells. Nature **363**, 71–74 (1993)
11. T. Kurahashi, A. Menini, Mechanism of odourant adaptation in the olfactory receptor cell. Nature **385**, 725–729 (1997)
12. S. Firestein, How the olfactory system makes sense of scents. Nature **413**, 211–218 (2001). www.nature.com
13. K.C. Persaud, D.H. Lee, H.G. Byun, *Objective Odour measurements* (Byun University of Manchester Institute of Science and Technology, Manchester, 2002)
14. ASTM E679-04: Standard practice for determination of Odour and taste threshold by a forced-choice ascending concentration series method of limits
15. ASTM E544-99, Standard practice for referencing supra-threshold Odour intensity EN13725: 2003: Air Quality—Determination of Odour concentration by dynamic Olfactometry, 2004
16. M. Corradi, P. Gergelova, A. Mutti, Exhaled volatile organic compounds in non respiratory diseases. European respiratory monograph; ISSN: 1025-448x. doi:10.1183/1025448.x00018809
17. http://www.cf.ac.uk/biosi/staffinfo/jacob/index.html

Chapter 4
Odor

4.1 Introduction

The human nose is a sense organ that senses odors. Experiencing smell sensation consists of the initial impacting of vapors on the sensor part of the nose. The sense of smell has long played a fundamental role in human development and biosocial interactions. Consequently, the olfactory sense has become a key element in the development of many commercial industries that manipulate the aroma properties of their manufactured goods in order to improve product appeal, quality, and consistency so that consumers quickly identify with individual brands having unique scents.

A more flexible way of presenting the primary odors to clarify the idea of complex odors is through the use of Henning's odor Prism [1] (Fig. 4.1). Six primary odors are located at the corners of the prism. All other odors are mixtures of the primary odors and located on the surfaces and edges of the prism. Thus, odors consisting of two components would be represented on the edges of the prism, three component mixtures occupy the triangular surfaces, and four-component mixtures occupy the square surfaces.

Odor can be defined as the "perception of smell" or in scientific terms as "a sensation resulting from the reception of stimulus by the olfactory sensory system". Whether pleasant or unpleasant, odor is induced by inhaling air-borne volatile organics or inorganic. The sources of smell are volatile compounds of low molecular weight (molecular weight $<$300 ppm) that are actually recognized as odor. The estimated total number of odor-producing materials is around several hundred thousand kinds of low molecular weight volatile compounds.

Materials that humans can perceive as odor include many organic compounds, such as the aromatic series (benzenes), as well as compounds containing oxygen (alcohol, aldehyde, ketone, and fatty acid), sulfur, and nitrogen, among others. Among inorganic substances, fluorine, chlorine, bromine, sulfur, hydrogen sulfide, nitrogen dioxide, and ammonia, etc., all possess objectionable odors [2].

H. K. Patel, *The Electronic Nose: Artificial Olfaction Technology*,
Biological and Medical Physics, Biomedical Engineering,
DOI: 10.1007/978-81-322-1548-6_4,

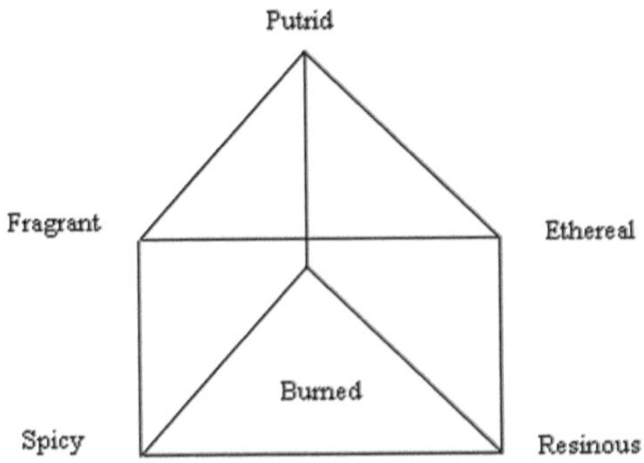

Fig. 4.1 Henning's odor prism

4.2 Odor Perception

The chemical senses, for smell (olfaction) and taste (gustation), are mainly measured to be the oldest ones in evolutionary development. While humans are a comparatively recent development in evolutionary terms, the function of our sense of smell is the same as for other type: it helps us to evaluate our environment. In easy talking of behavior, perception of odors can lead to two basic behavioral responses: avoidance or approach. These responses can occur for example in judging food or water, but also in a social context.

The human sense of smell helps us to evaluate our environment in a very straight manner. The sensor in the nose opening is a direct interface between the brain and the environment. It is a highly sophisticated sense, which interacts with our life and behavior on many levels. The process of odor detection, perception, and evaluation is therefore understandably complex.

A human has an ability to sense and differentiate up to 3,000 odors. Latest research indicates that as many as 1,000 genes out of the total of 100,000 in our genome are dedicated to sense of smell. This significant proportion of 1 % suggests that the sense of smell is of considerable importance in evolutionary terms [3].

The sense of smell is closely related to long-term memory. The nerves that connect the sensor to the brain lead directly to the hippocampus, which is the part of the brain that regulates basic functions, such as the organization of long-term memory and emotions. It is, therefore, not surprising that smells are often highly associative and can elicit vivid memories of experiences that occurred even in early childhood. This associative aspect is highly relevant to environmental odors. Once a negative association is formed, it is very difficult to change the appreciation of that particular odor stimulus in an individual. This helps to explain why an odor problem from the past often seems to haunt site operators, even after the odor emissions have been significantly reduced.

Smell is confined of residual molecules in the air by receptor cells and the resulting perceptions of 'smell' tell the living being about something that is nearby. Smell also has an intense effect on our day-to-day experience and can trigger memories.

From person-to-person, senses are variable in their range and acuity. Through comparison with others, it was recognizing that some people can see or hear what most of other miss, or others miss what most of us see or hear.

Odor, which refers to smells, can be utilized as a marker to identify certain problems or sources of interest. These include air pollution, environmental contamination, disease diagnostics, and as noted above, human identification in crime investigations. Odor consists of volatile organic compounds (VOCs) that typically have relative molecular masses between 30 and 300 g/mole [4]. Heavier molecules do not occur as VOCs because they generally have a vapor pressure at room temperature that is too low to be active odorants. The volatility of molecules is determined by both their molecular weight and their intermolecular interaction, with non-polar molecules in general being more volatile than polar ones. As a consequence the most odorous molecules tend to have one or two polar functional groups. More functional groups in general result in molecules that are much less volatile.

The perception of smell consists not only of the feeling of the odors themselves but of the experiences and emotions associated with these sensations. Smells can evoke strong emotional reactions. In surveys on reactions to odors, responses show that many of our olfactory likes and dislikes are based purely on emotional associations.

4.3 Molecular Biology of Smell

Smells are with us all the time. Although the human sense of smell is lacking compared to many animals, it is still very acute. We can recognize thousands of different smells, and we are able to detect odors even in infinitesimal quantities. All living organisms including humans respond to substances in their environment.

Humans have three distinct chemical senses which constitute the perception of flavor. These are the sense of taste, the sense of smell (olfaction), and trigeminal sense (irritation). The most significant contribution is made by the sense of smell. In addition, the sense of taste and trigeminal sense are much simpler systems than smell. The sense of smell has a much broader range and more power of classification than either taste or trigeminal sense. As a result, the sense of smell is crucial for humans and animals. It is considered one of the most ancient of senses. Smell allows organisms with olfactory receptors to identify food, mates, predators, and also provides not only sensual pleasure such as the odor of flowers, but also warnings (e.g., spoiled food, chemical dangers). Therefore, it is one of the most important systems of living organisms.

Our organs of smell consist of two patches (made up of about five or six million yellowish cells—high up in the nasal passages) of epithelial tissue not much larger than a postage stamp located near the top of the nasal cavity. It contains millions of olfactory receptor cells, bipolar neurons ending in olfactory cilia. The cilia (nonmotile) are cellular extensions bathed in mucus to which chemicals bind.

These cells are surrounded by supporting cells. At the base of the epithelium are basal cells. Bundles of axons of the olfactory receptor cells pass through the foramina of the cribriform plate of the ethmoid bone and then synapse with neurons of the olfactory bulbs of cranial nerve I.

The actual breath contains mixtures of oxygen, carbon dioxide, water vapor, nitrogen, inert gases, in addition may also contain various elements and more than 1,000 trace volatile compounds. Concentrations ranges are from ppm to ppt. The Volatile Organic Compounds (VOCs) commonly found in normal breath are acetone, ethane, and isoprene which are nothing but the metabolic products [5].

Approximately a thousand different kinds of olfactory receptors are there in human nose, and it is thought that each can sense a single kind of chemical bond in a molecule.

4.4 Types of Odor

Seven primary odorant classes are listed in Table 4.1

4.4.1 Characteristics of Odor

- Substances of similar or dissimilar chemical constitution may have similar odor. Nature and strength of odor may change on dilution.
- Weak odor is not perceived in presence of strong odor.
- Odor of same strength blends to produce a combination in which one or both may be unrecognizable.
- Constant intensity of odor causes an individual to quickly loose awareness of the sensation and only noticed when it varies in intensity.
- Fatigue for one odor may not affect the perception of dissimilar odor but will interfere with the perception of similar odor.
- An unfamiliar odor is more likely to cause complaint than a familiar one.
- Two or more odorous substances may cancel the smell of each other.
- Odor travels downwind.

Table 4.1 Primary odorant

Odor	Example
Camphoric	Mothballs
Musky	Perfume/Aftershave
Roses	Floral
Peppermint	Mint gum
Ethereal	Dry cleaning fluid
Pungent	Vinegar
Putrid	Rotten Eggs

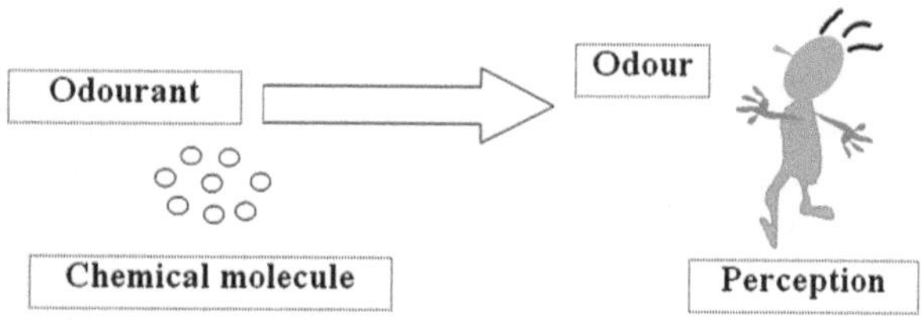

Fig. 4.2 Chemical molecule odorants lead to the perception of odor

- Person can smell at a distance.
- Many animals have keener sense of olfaction than man.
- Likes and dislikes often depend on association of the scent with pleasant or unpleasant experiences.

To smell something the chemical must first become aerosolized, it must be volatile. It must then dissolve in the mucus before binding to the olfactory cilia. The chemical binding causes the bipolar neurons to depolarize. There are many (at least a thousand) different receptor proteins on the membranes of the olfactory cilia. A receptor protein may respond to several different chemicals and each chemical may bind to several different kinds of proteins. The average person can detect about 4,000 different odors. The odor is the perception experienced when one or more chemical substances in the air come in contact with the various human sensory systems (odor is a human response) [6], see Fig. 4.2. Some odors may be composed of one key chemical compound (one odorant); however, most often, odors are a complex mixture of various chemicals (many odorants) [6].

4.5 Odor Parameters

Odors can be quantified by five parameters that outline the human response. These parameters include: odor thresholds, odor intensity, odor persistency, hedonic tone, and odor characterization.

4.5.1 Odor Detection Threshold

It is the lowest odorant concentration necessary for detection by a certain percentage of the population, normally 50 %. This concentration is defined as 1 odor unit.

Odor detection capability or threshold or concentration is a sensory property refers to the minimum concentration that produces an olfactory response. With odor intensity at or just above "threshold" odor become difficult to perceive.

An **odor detection threshold** relates to the minimum odorant concentration required to perceive the existence of the stimulus, whereas an **odor recognition threshold** relates to the minimum odorant concentration required to identify the stimulus. The detection threshold occurs at a lower concentration than the recognition threshold.

Odor concentration is measured as dilution ratios and reported as Dilution Threshold and Recognition Threshold or Dilution to Threshold (D/T) and sometimes assign the pseudo-dimension of odor units per cubic meter. Dilution to Threshold (D/T) ratio is a measure of the number of dilutions needed to make the odorous air nondetectable. Odor unit is the concentration divided by the threshold.

4.5.2 Odor Intensity

It is the perceived strength of an odor above its threshold. It is related to the odorant concentration. It is determined by an odor panel and is described in categories which progress from “not perceptible”, then “very weak”, through to “extremely strong” [7].

Generally odor intensity increases with the odorant concentration. The relationship between intensity and concentration can be expressed as:

$$I = k(C)^n$$
$$\text{or } \mathrm{Log}\, I = \mathrm{Log}\, K + n\,\mathrm{Log}\,(C)$$

where,

I Intensity;
C Concentration;
K Constant; and
n Exponent.

This is known as Stevens’ law or the power law [8]. Odors are ranges (n) from about 0.2–0.8, depending on the odorant. For an odorant $n = 0.2$, a tenfold reduction in concentration decreases the perceived intensity by a factor of only 1.6, whereas for an odorant with $n = 0.8$, a tenfold reduction in concentration lowers the perceived intensity by a factor of 6.3. This is an important concept that is related to the basic problem of reducing the odor intensity of a substance by air dilution or other means. Odor Intensity is expressed in ppm of butanol.

The German standard Olfactometry Determination of Odor Intensity VDI 3882 Part 1 (VDI, 1992) provides qualitative descriptions of odor intensity with a numerical scale that may be used in back-calculating the corresponding odor concentration. These descriptions are represented in Table 4.2.

Odor intensity is a useful dimension to compute because some odors are perceived as being stronger than others. In other words, all odors will be just detectable at a concentration of 1 OU/m^3, however, at twice the concentration, or 2 OU/m^3, some odors may be perceived as very weak while others may be perceived as distinct. At ten times the concentration, or 10 OU/m^3, one odor may be perceived as distinct while another odor at 10 OU/m^3 concentrations may be very

Table 4.2 Odor intensity categories

Odor strength	Intensity level
Extremely strong	6
Very strong	5
Strong	4
Distinct	3
Weak	2
Very weak	1
Not perceptible	0

strong. This means that defining an odor criterion based on odor concentration, as has historically been done for the purposes of managing odor impact on the community will result in different perceived odor strengths. The only time this will not occur is when the odor criterion is equal to the detection threshold (i.e., 1 odor unit) which effectively becomes a “no impact” criterion.

The odor intensity result is expressed in ppm (PPM) of butanol (n-butanol) [8]. A larger value of butanol means a stronger odor, however, not in a simple numerical proportion. The butanol concentrations provide a referencing scale for documentation and communication in a reproducible format [9]. Common butanol intensity referencing scales include:

- 12-point static scale starting at 10 ppm butanol with a geometric progression of two,
- 10-point static scale starting at 12 ppm with a geometric progression of two,
- 8-point dynamic scale starting at 12 ppm with a geometric progression of two, and
- 5-point static scale starting at 25 ppm with a geometric progression of three.

4.5.3 Scaling Methods

There are various methods to scale perceived magnitude, a category scale, which can be either number- or word-categorized, is commonly used. Numerical values on this scale do not reflect ratio relations among odor magnitudes (e.g., a value of 2 does not represent a perceived magnitude twice as great as a value of 1).

Table 4.3 gives examples of category scales. Although category scaling procedures can be advantageous in field situations, ratio scaling techniques are used frequently in the laboratory [10]. Ratio scaling procedures require observers to assign numbers proportional to perceived magnitude. For example, if the observer is instructed to assign the number 10 to one concentration and a subsequently presented concentration seems three times as strong, the observer calls it 30; if another seems one-half as strong, the observer assigns it 5. This ratio scaling procedure, called magnitude estimation, was used to derive the power function for butanol (Fig. 4.3). Ratio scaling techniques are allowed for such relationships because they require subjects to produce numbers to match perceived sensations in which the numbers emitted reflect the ratio relations among the sensations.

Table 4.3 Examples of category scales [11]

Number category		Word category	
Scale 1	Scale 2	Scale 1	Scale 2
0	0	None	None at all
1	1	Threshold	Just detectable
2	2.5	Very slight	Very mild
3	5	Slight	Mild
4	7.5	Slight-moderate	Mild-distinct
5	10	Moderate	Distinct
6	12.5	Moderate-strong	Distinct-strong
7	15	Strong	Strong

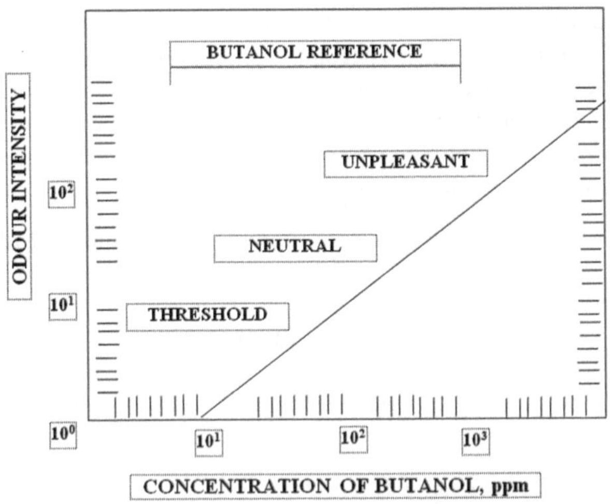

Fig. 4.3 Standardized function relating perceived magnitude to concentration of 1-Butanol [10]

A hybrid of category and ratio scales is known as the labeled magnitude scale [12]. This scale is proposed to acquiesce ratio-level data with a true zero and an orderly relationship among the scale values, such that any stimulus can be expressed as being proportionately more or less intense than another. Because it allows subjects to use natural language descriptors to scale perceived experience, it often requires less training than ratio scales and produces absolute intensity estimates of perceived sensation [13] (Fig. 4.4).

To measure odor intensity another, the fourth way is to match the "intensity of odorants". An observer can be given a concentration series of a matching odorant (e.g., 1-butanol) to choose the member that matches most closely the intensity of an *unknown* odorant. The matching odorant can be generated by a relatively inexpensive olfactometer is shown in Fig. 4.5.

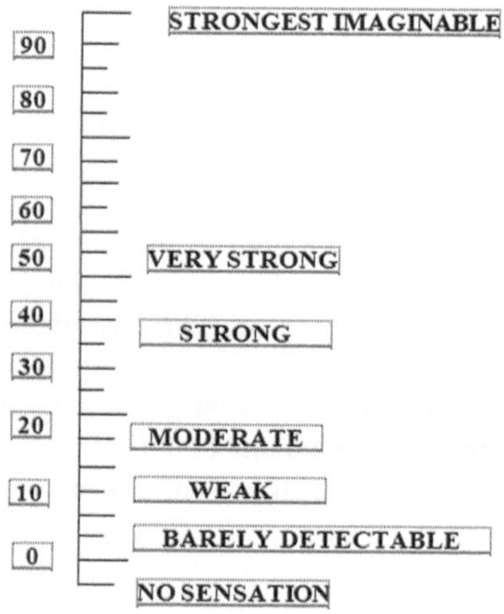

Fig. 4.4 Labeled magnitude scale

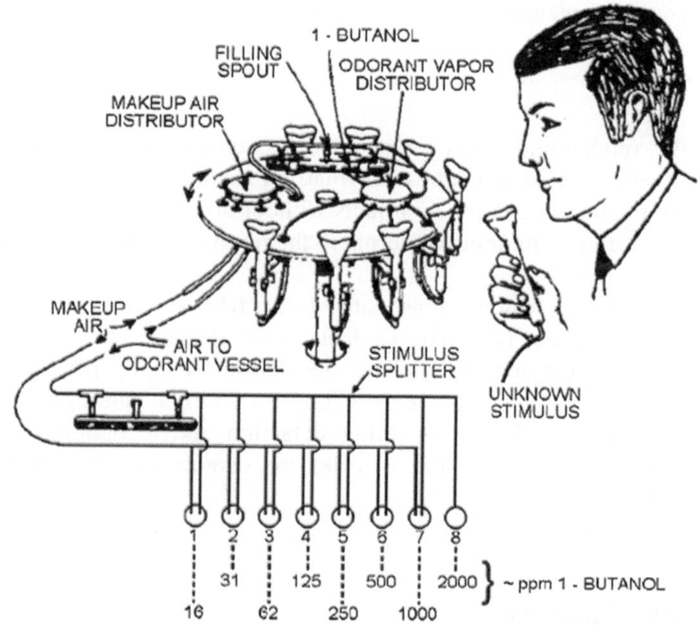

Fig. 4.5 Panelist using Dravnieks binary dilution olfactometer [14]

4.5.4 Hedonic Tone

Hedonic tone is a measure of the pleasantness or unpleasantness of an odor. It is the degree to which an odor is perceived as pleasant or unpleasant. Such perceptions differ widely from person-to-person, and are strongly influenced, inter alia, by previous experience and emotions at the time of odor perception.

An evaluator uses her/his personal experience and memories of odors as a hedonic tone referencing scale. During training, evaluator becomes aware of their individual odor experience and memory referencing. The reported hedonic tone value by an odor testing laboratory is an average of individual hedonic tone values assigned by each evaluator [15]. The members of a panel of evaluators are asked to indicate perceived hedonic tone at each presentation as a value from the nine-point hedonic tone scale:

+4 very pleasant
+3 pleasant
+2 moderately pleasant
+1 mildly pleasant
0 neutral odor/no odor
−1 mildly unpleasant
−2 moderately unpleasant
−3 unpleasant
−4 offensive.

Subjective: relying upon ones personal feelings or beliefs; relating to or arising within one's self or mind in contrast to what is outside.

Objective: treating or dealing with facts without distortion by personal feelings or prejudices; dealing with things external to the mind rather than with thoughts or feelings.

The assigning of a hedonic tone value to an odor by an evaluator is "subjective" to the evaluator [16]. The evaluator's experiences and memories force their personal feelings and beliefs into the decision-making process. Through training, evaluators assign a hedonic tone and then set aside their personal feelings and make objective decisions regarding detection and recognition thresholds, intensity referencing using a butanol scale, and character identification using a category reference [8].

4.5.5 Odor Character

It is basically what the odor smells like. It allows differentiating between various types of odors. For example, Nitrogen dioxide (NO_2) gas has a biting and irritating odor and it is an air pollutant. The character of an odor can change with intensity. The quality or character of an odor is difficult to evaluate. Main difficulty is that odors can differ along many dimensions. One way to evaluate quality is to ask panelists for judge similarity between a test sample and various reference samples, using a five-point category scale. For certain applications, reference odorants can be chosen to represent only that portion of the qualitative range relevant to the odor problem under investigation (e.g., animal odors). Another procedure is to ask panelists to assess the degree of association between a test sample's quality and certain verbal descriptors (e.g., sweaty, woody, chalky, sour). An odor can be characterized either by an open-ended word description or by multidimensional scaling. Multidimensional scaling is based on similarity and dissimilarity judgments in comparison to a set of standard odors or to various descriptors. In some instances, the interest may be merely whether an odor's quality has changed as a result of some treatment (e.g., use of a bacteriostat). Under these circumstances, samples of air taken before and after treatment can be compared directly (using a simple scale of similarity) or indirectly (with appropriate verbal descriptors).

4.5.6 Odor Unit

Odor concentration can be expressed as the number of unit volumes that a unit volume of odorous sample occupies when diluted to the odor threshold with non-odorous air. If a sample of odorous air can be reduced to threshold by a tenfold dilution with pure air, the concentration of the original sample is said to be 10 odor units. Hence, odor units are equivalent to multiples of threshold concentrations. Odor units are not units of perceived magnitude. The "odor unit" seems to be the most common index for odor emission control. A number of states in the US have a source emission standard. However, there are problems in using the odor unit as a standard: (1) because of the variability of people, who serve as the detectors for generation of the odor unit, data vary from laboratory-to-laboratory, and (2) the odor unit includes *no measure of the importance of the odor.*

The European Odor Unit (OUE) is defined in terms of N-butanol.

To calculate the European Odor Units [17]:

1. Determine concentration of n-butanol at its Odor Detection Threshold (ODT_b). This is the Odor Detection Concentration for n-butanol (ODC_b).
2. Determine the Odor Units for the "mixed sample": this is the Odor Detection Threshold of the unknown sample adjusted to the Odor Detection Concentration for n-butanol

OUE = (ODT X ODC_b)/40 ppb
OUE = European Odor Units
ODT = Odor detection threshold (ratio) of the sample
ODCb = Odor concentration of n-butanol at its detection threshold
40 ppb = the "definition" of 1 OUE in terms of n-butanol

European standards require that Odor Detection Concentration (ODC_b) between 20 and 80 ppb for each panelist, thus panelists are screened prior to their participation in an olfactometry panel. One "European Odor Unit" is 123 mg n-butanol (40 ppb) by definition so, if the Odor Detection Concentration (ODC_b) determination to be other than 40, it must adjust as per ODT accordingly.

4.6 Effects of Odor

Odor affects human beings in a number of ways. Strong, unpleasant, or offensive smells can interfere with a person's enjoyment of life especially if they are frequent and/or persistent. Major factors relevant to perceived odor nuisance are:

- Offensiveness
- Duration of exposure to odor
- Frequency of odor occurrence
- Tolerance and expectation of the receptor.

Though foul odor may not cause direct damage to health, toxic stimulants of odor may cause ill health or respiratory symptoms. Secondary effects, in some, may be nausea, insomnia, and discomfort. Very strong odor can result in nasal irritation; trigger symptoms in individuals with breathing problems or asthma. On the

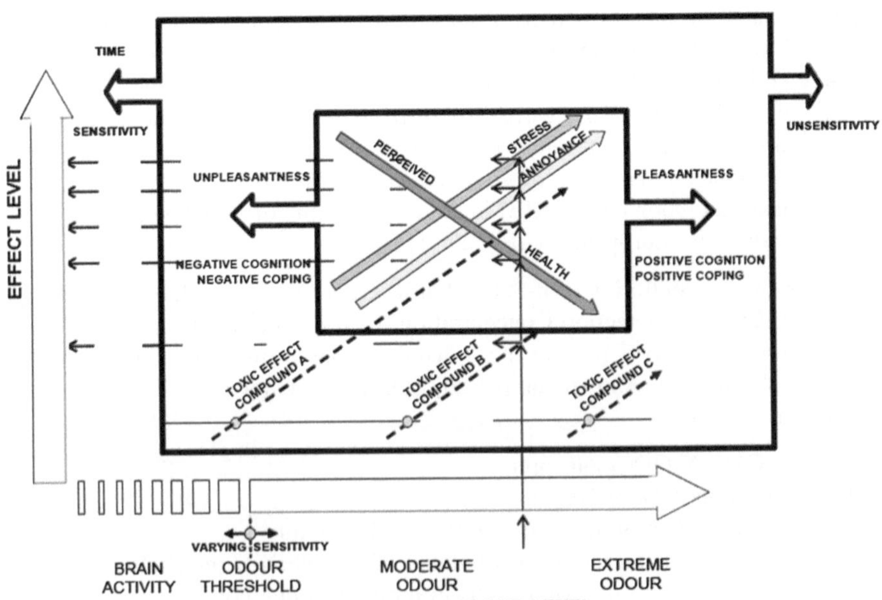

Fig. 4.6 Schematic illustration of health-related response to odor and odorants [18]

Table 4.4 Methods used for elimination of Odor

1	Physical	Aqueous cleaning Air ventilation Absorption to activated charcoal
2	Chemical	Combustion Neutralization oxidation Chemical absorption
3	Biological	Microorganism Antimicrobial agent
4	Sensory	Masking with air freshener

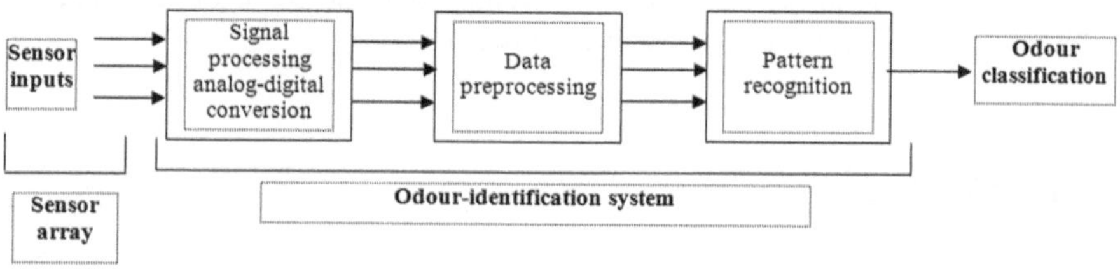

Fig. 4.7 An odor classification system

economic front, loss of property value near odor causing operations/industries and odorous environment is partly a consequence of offensive odor. Figure 4.6 shows the relationship between odor and human reactions related to well-being and health. Table 4.4 list the methods used for eliminating unpleasant/hazardous odor (Fig. 4.7).

4.7 Odor Objective and Classifications

The signals are obtained by measuring the change in the sensor's electrical properties upon exposure to a gas. Next, the system converts these signals to digital form, reduces data complexity in the preprocessing stage, and then employs pattern recognition to classify the odor.

There are three basic ways to classify odors.

- By likes or dislikes
- By associations
- By chemical structures.

Likes and dislikes are specific because tastes differ. Association becomes complicated because associations are different and odorant possesses different qualities. Chemical classes or structures are very interesting ways of classifying odors. They are grouped in families according to molecule structure.

Table 4.5 Relationship between odor intensity and odor index

Odor intensity	Range of odor index
2.5	10–15
3.0	12–18
3.5	14–21

Odor classification is studied using psychophysical methods would suggest that there are several primary odor groups. These include camphor-like, musky, floral, pepperminty, ether-like, pungent, and putrid. Unique mixtures of aromatic compounds including phenolics, ketones, and terpenes [5] determine the characteristic odors of fruits and vegetables. Smell sensitivity is genetically determined and varies from person-to-person, hence different people perceive or may react to different odors, or react differently to the taste of foods.

4.7.1 Odor Index

The odor index is defined as the dimensionless ratio of the vapor pressure and the 100 % odor recognition threshold (the concentration at which 100 % of the odor panel detect/recognized the odor as being representative of the odorant being studied). The odor index provides information on the potential of a particular compound to cause odor problems under evaporative conditions [19].

The "Odor Index" is used in Japan to quantify the intensity of odors [19]. The odor index is equal to ten times the log of the odor concentration (i.e., Odor Index $= 10 \times \log$(Odor Concentration). Table 4.5 shows the relationship between odor intensity and odor index. The odor concentration is measured using the Triangular Odor Bag Method. The odor indices of the common odorous compounds are given in Table 4.6. Also the variance of odor measurement illustrates in Fig. 4.8.

4.8 Odor Analysis Techniques

Conflicting criteria must be considered when choosing an odor assessment.

For example:

- Timing of Data Collection—Must data be collected on-line? If not, how soon a sample is collected, should the analytical results be available?
- Sensitivity, Selectivity, and Stability—What range of analyte concentration or property value can be expected? What are the possible interferences or matrix effects? How important is day-to-day reproducibility? How important is instrument-to-instrument reproducibility?

Table 4.6 Odor indices of the common odorous compounds [20]

Compound	Odor index
Acetaldehyde	4,300,000
Acetic acid	15,000
Acetic andydride	12,800
Acetone	720
Acrylic acid	4,210
Allylalcohol	13,800
Allylchloride	17,900
Ammonia	167,300
Benzaldehyde	22,000
Benzene	300
Benzylchloride	28,000
1,3 butadiene	2,530
n-butane	480
n-butanol	120
Sec. Butanol	400
Tert. Butanol	55,900
n-butylacetate	1,200
n-butylamine	395,000
n-butylchloride	6,300
n-butylether	13,400
n-butylmercapton	49,340,000
n-butysulfide	658,000
n-butyric acid	50,000
Carbon disulfide	44,430
Carbon tetrachloride	540
Chlorobenzene	52,600
Chlorobromomethane	350
Chloroform	70
Diethylamine	880,000
Diethylsulfide	14,400,000
Diethylketone	1,900
Dimethylamine	280,000
Dimethylsulfide	2,760,000
Ethane	25,300
Ethanol	11
Ethylacetate	1,900
Ethylene	57,100
Ethylether	1,939,000
Formaldehyde	5,000,000
Hydrogen sulfide	17,000,000
Isobutane	3,000,000
Isobutanol	320
Isobutene	4,640,000
Isobutylacetate	3,300
Methanol	22

(continued)

Table 4.6 (continued)

Compound	Odor index
Methylacetate	1,100
Methylamine	940,000
Methylamylalcohol	12,650
Methylchloride	200,000
Phenol	16
Propane	425
Propionic acid	112,300
Skatole	30,000
Toluene	720
Valeric acid	256,300

- Cost and Benefit of Analysis—What is the total cost of running the instrument including maintenance, methods development, sample handling, operator training, and disposables?

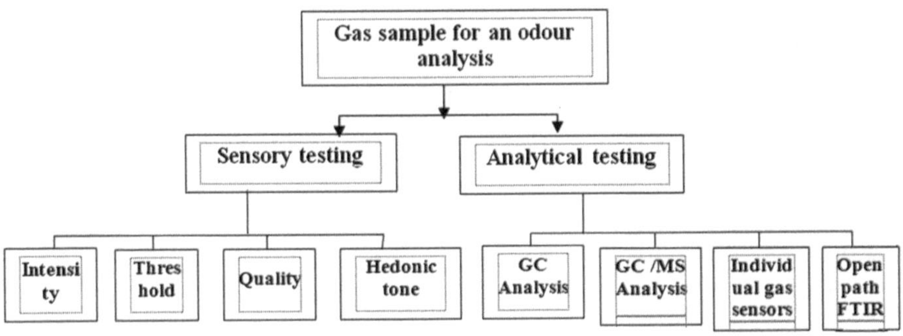

Applications of odor measurement in industries

- Foods: Assessment of freshness, quality control of foods, and beverages
- Chemicals: Quality monitoring of synthetic chemicals
- Odor abatement facilities: Efficiency check of deodorizing facilities
- Deodorants: Deodorizing efficiencies
- Air cleaners: Efficiencies of deodorization
- Air conditioners: Measurements of malodors from air conditioners
- Refrigerators: Deodorizing system of offensive odors in a refrigerator
- Vacuum cleaners: Treatments for bad smell of exhaust air
- Automobiles: Odors in a new or old cabin, malodors from stuff.

Odor evaluation methods could be categorized from various standpoints. Figure 4.9 shows same and also Table 4.7 categorized odor measurement technology with remarkable advantages and disadvantages.

The sense of odor involves three principal factors: concentration of odorous substances, sensory odor intensity, and odor quality including hedonic tone.

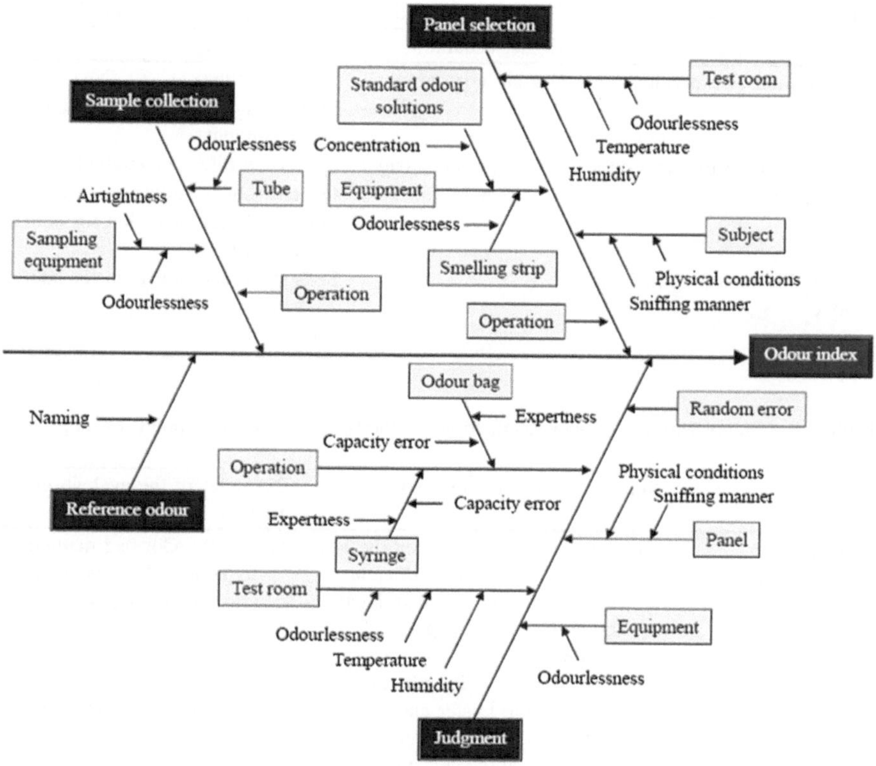

Fig. 4.8 Variance of odor index measurement [17]

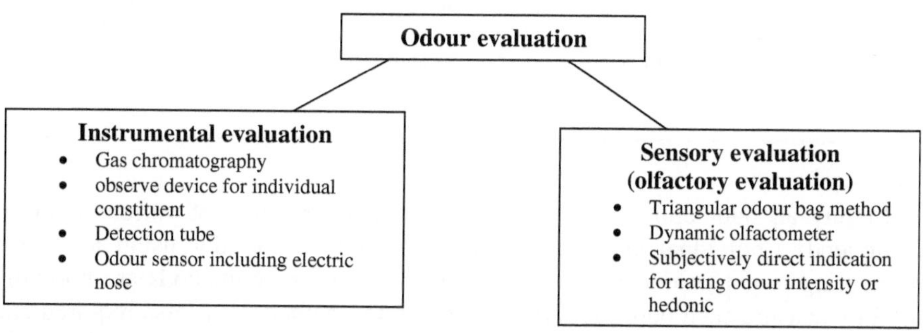

Fig. 4.9 General categorization of odor evaluation method

Thus, the odor evaluation method can be categorized from these three aspects as shown in Table 4.7 [21].

On the other hand, it is possible to divide odor evaluation methods into three categories, as shown in Table 4.8 [21], from the viewpoint of application purpose.

Table 4.7 Categorization of odor evaluation methods from the viewpoint of indicator objectives

Objective of indicator	Higher accuracy	Lower accuracy
Concentration of Individual constituent	Gas chromatography	Detection tube Monitoring device for specified constituents
Odor index, odor unit and odor intensity	Triangular odor bag method	Simplified olfactory methods Odor sensor including electric nose Sensory direct indication for odor intensity
Odor quality including hedonic tone		Sensory direct indication for odor hedonic tone Electric nose

Table 4.8 Categorization of odor evaluation methods from the viewpoint of application purposes

Category	Summarize method analysis	Examples of the evaluation method
Evaluation for compliance	Obligatory method based on the law. Necessary to ensure sufficient accuracy for judgment in legal action. Generally expensive method	Triangular odor bag method, dynamic olfactometer, gas chromatography, etc
Evaluation for voluntary management	Not obligatory. Use in voluntary management. Easier and cheaper methods are better. Highly accurate methods are not always necessary. Highly frequent monitoring is possible	Detection tube, odor sensor, subjectively direct rating for intensity or hedonic tone, etc
Evaluation for characterization of odor emission source	Characterization of time-dependent change of odor emission from the odor source, detailed composition of odorous constituents and contribution of constituents to sensory magnitude of odor	Continuous monitoring using odor sensor, GC olfactometer, GC/MS, electric nose

Mainly an odor emission consists of a complex mixture of various odorous compounds. Individual chemical compounds analysis of odor using analytical monitoring is not practical. Because of that, odor sensory methods are normally used to measure such odor instead of instrumental methods. It is also required and useful for monitoring odor from: source of odor emissions and in the ambient air. These two different conditions required different approaches for measuring odor. Also odor samples collection is easily available for a source emission than for an odor in the ambient air. Note that due to atmospheric dilution, the odor in the ambient air is typically much lower in intensity than it is at source. Thus the sensitivity of the odor sensory method must be significantly greater for measuring ambient odor than for source odor emissions.

For identified composite, the odor strength can be reliably estimated by measuring the concentration of the chemical, but, for mixtures of unknown substances, sensory method is preferred.

Measurement of odor is identical for source emission air samples and for ambient air at the property line and in the community. For determining criteria of odor limits three basic odor limits built-in for conformity:

1. Ambient odor concentration limits
2. Ambient odor intensity limits
3. Source emission odor concentration limits.

Measuring odor can be accomplished in a number of ways: instrumental methods/chemical analysis, electronic methods, and sensory test methods/olfactometry.

4.8.1 Instrumental Methods

Mainly instrumental methods available depend on the application of gas chromatography (GC), including gas chromatography-mass spectrometry (GC–MS), since this prime separation technology is competent to the efficient separation required for analysis of complex mixtures of odor. A mixture of volatile substances is injected into a column of gas chromatography, which separates the compounds based on their relative vapor pressures and polarities. The compounds are then detected as highest point, which have specific maintenance times and maximum areas, which can be used for qualitative and quantitative determinations, respectively [20].

4.8.2 Sensory Methods (Olfactometry)

Measuring odor using the human sense of smell (Olfactometry) is the most valid way of measuring odor. Accurate measurement of odorous compounds and odor impact have been challenging because these compounds acquire generally varying physical and chemical properties and are present at concentrations ranging from high parts per million (ppm) to low parts per billion (ppb). Olfactometry is the most commonly used method to measure the concentration of odor in air. Olfactometry is carried out using an instrument called an olfactometer.

Furthermore, each odorant has a unique odor and odor detection threshold which means that compounds, even if present at the same concentration, may have markedly different odor impacts. The sensory methods used for measuring odor level include [20]:

- Static dilution olfactometry
- Dynamic dilution Olfactometry
  - free choice (Yes/No Olfactometer)
  - forced Choice Olfactometer
  - Triangular Forced Choice Olfactometer.

(a) Static Dilution Olfactometry

In Static dilution a sample of odorous gas is diluted in 100 ml glass syringe at various dilution levels. The diluted samples are expelled into the nostrils of the panelists. The odor detection threshold is determined graphically from the dilution levels and panelist response data. A panel of eight panelists is preferred.

(The American Society for Testing and Materials (ASTM D-1391) static dilution/syringe method was developed in 1978 and was withdrawn by the ASTM E-18 Committee on March 29, 1986, however, the procedure is still in use, principally in the USA).

(b) Dynamic Dilution Olfactometry

A dynamic olfactometer provides a continuous and constant-diluted odor stimulus by mixing controlled flows of sample and odorless air. With dynamic methods, larger samples are used and dilutions are presented at more reproducible flow rates and for longer duration for panelists to evaluate. The presentation of odorous sample dilutions to panelists and their responses depend on three sensory effects: Judgment criterion, anticipation and adaptation. The judgment criterion determines how the panelist is to response when asked whether or not an odor is sensed. This is the case particularly when a single stimulus is presented and a yes or no answer is requested as to the sensation of odor. The anticipation effect is a tendency to expect an odor to occur when odor less or weak samples are consecutively presented. The adaptation effect is a temporary loss of sensitivity after smelling an odor. When a weak odor is detected initially, the same odor may not be detected again after smelling a stronger odor unless the panelist has had sufficient time to recover his or her olfactory sense.

When a forced choice method is used, a panelist, typically trained to conduct these evaluations, must identify the presentation that is different from the others at each level, even if it is a guess. This permits use of all the data. The threshold of detection is the dilution level at which the panelist can determine a difference between the diluted and the odorless samples. After the detection threshold is reached, the panelist continues the evaluation at the next level or two to be certain the identification was not made by chance. Examples of the dilution to threshold methods include use of scentometery and Olfactometry. Figure 4.10 illustrates the forced choice Olfactometer and Fig. 4.11 illustrates the dynamic olfactory schematic diagram.

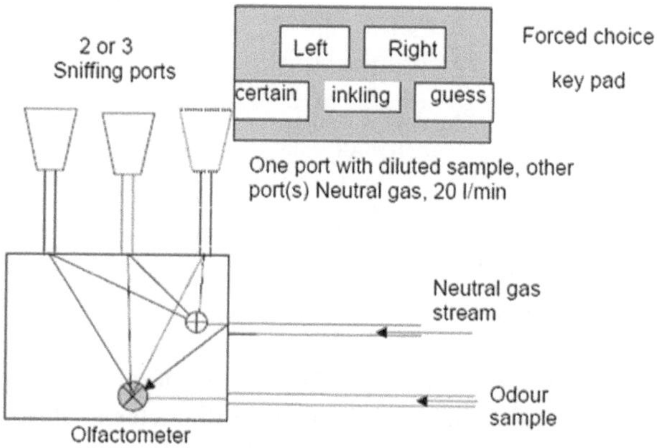

Fig. 4.10 Schematic diagram of a forced choice olfactometer [22]

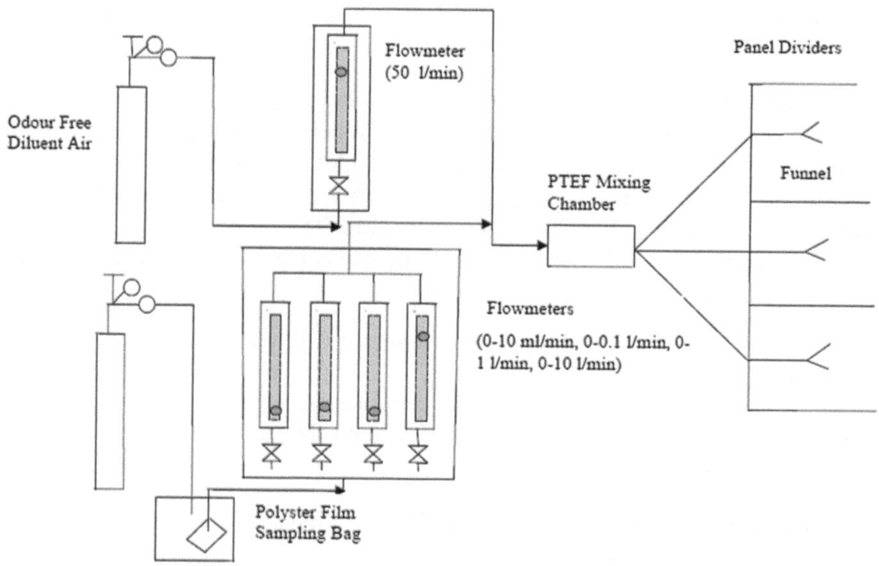

Fig. 4.11 Dynamic olfactory [20]

4.8.3 Scentometery

One method of odor concentration evaluation that is available on-site employs the use of a Scentometer® (Barneby and Cheney, Columbus, OH) or a Nasal Ranger® (St. Croix Sensor y, St. Elmo, MN). The Scentometer® is a plastic box with a number of air inlets and two sniffing ports. Two of the air inlets have activated charcoal filters to remove odors and provide clean air. The remaining inlets are of

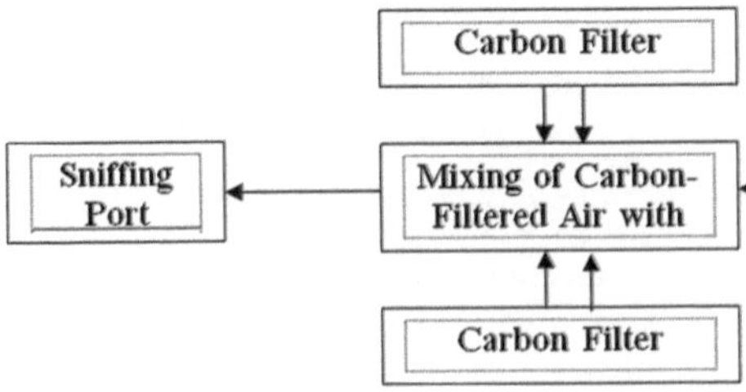

Fig. 4.12 Block diagram of field olfactometer air flow

varying diameter to permit a range of dilutions of odorous air to be sampled. An observer begins by opening the port of smallest diameter to start with the largest dilution (lowest concentration) of the odor [23,1]. An air flow diagram in field Olfactometer is shown in Fig. 4.12.

The method of calculating Dilution to Threshold (D/T) for a field Olfactometer is:

$$\begin{aligned}\text{Dilution Factor} &= \text{Volume of Carbon Filtered Air}/\ \text{Volume of Odorous Air}\\ &= \mathbf{D/T}\end{aligned}$$

As successively larger ports are opened, the dilution of the odorous air decreases and the odor concentration increases. When the evaluator can first detect the odor, the odor threshold has been reached. Odor concentrations are expressed as dilutions to threshold. The range of dilutions to threshold possible for the Scentometer includes 1.5, 2, 7, 15, 31, 170, and 350.

Field olfactometry has the following key advantages over laboratory olfactometry for measurement of ambient odors.

1. Lower method detection limit (most laboratory olfactometers have a method detection limit of 5–10 dilutions);
2. Immediate results (Laboratory results can take 1–5 days to receive a report);
3. Eliminates concern for deterioration of odor in the sample bag; and
4. Low per sample cost.

Field olfactometry with a calibrated field olfactometer is a cost effective means to measure odor strength. Facility operators, community inspectors, and neighborhood citizens can confidently measure odor strength at specific locations around a facility's property line and within the community when using a calibrated field olfactometer.

4.8.4 Olfactometry

Olfactometers operate much like the Scentometer® and the Nasal Ranger®. The primary differences are olfactometers that are not portable and an operator closely controls sample delivery. Larger dilution to-threshold ranges are available.

During an odor test, the odor panelist (assessor) sniffs a diluted sample of the odor as it is discharged from the olfactometer as one of three sample presentations (one presentation with the diluted odor and two with odor free air). The assessor sniffs all three of the presentations and must select the one of the three that is different from the other two, even if they must guess. This statistical approach is called "triangular forced-choice." The assessor continues to additional levels of higher concentration (lower dilution) presentations following this procedure.

Therefore, "odor concentration" or odor strength is a number derived from the laboratory dilution of collected odors. The dilution ratio (total presentation volume divided by odor sample volume) at each sample presentation level is used to calculate the concentration of the evaluated sample.

The individual thresholds of eight to ten assessor responses are averaged to determine the detection threshold for which 50 % of individuals will observe the presence of an odor. The "detection threshold" value that is obtained from odor testing is derived from the dilution ratios, and is therefore dimensionless. However, the pseudo-dimensions of "Odor Units" (O.U.) or "Odor Units per Unit Volume" are commonly applied. For example: "Odor Units per cubic meter."

It should be noted that the dilution of the actual odor emission sample by the olfactometer is the physical process that occurs in the atmosphere downwind of the odor source. The "receptor" (citizen in the community) receives and sniffs the diluted odor. The dilution ratio is an estimate of the number of dilutions needed to make the actual odor emission "non-detectable" (Detection Threshold). If the receptor detects the odor, then the odor in the atmosphere is above the receptor's detection threshold level.

Appendix I: Odor Descriptors for Commonly Encountered Compounds [24–26]

Substance	Odor	Substance	Odor
Acetaldehyde	Apple, stimulant	Dimethyl sulfide	Rotten vegetable
Acetic acid	sour vinegar	Diphenylamine	Floral
Acetone	chemical/sweetish/solvent	Diphenyl sulfide	Burnt rubber
Acetonitrile	Ethereal	Ethanol	Pleasant, sweet
Acrylaldehyde	Burning fat	Ethyl acetate	Fragrant
Acrolein	Burnt sweet, pungent	Ethyl acrylate	Hot plastic, earthy
Acrylonitrile	Onion, garlic, pungent	Ethylbenzene	Aromatic
Aldehydes C9	Floral, waxy	Ethyl mercaptan	Garlic/onion, sewer, decayed cabbage, earthy
Aldehydes C10	Orange peel	Formaldehyde	Disinfectant, hay/straw-like, pungent
Allyl alcohol	Pungent, mustard-like	Furfuryl alcohol	Ethereal

(continued)

(continued)

Substance	Odor	Substance	Odor
Allyl chloride	Garlic onion pungent	n-Hexane	Solvent
Amines	Fishy, pungent	Hydrogen sulfide	Rotten eggs
Ammonia	Sharp, pungent odor	Indole	Excreta
Aniline	Pungent	Iodoform	Antiseptic
Benzene	Solvent	Methano	Medicinal, sweet
Benzaldehyde	Bitter almonds	Methyl ethyl ketone	Sweet
Benzyl acetate	Floral (jasmine), fruity	Methyl isobutyl ketone	Sweet
Benzyl chloride	Solvent	Methyl mercaptan	Skunk, sewer, rotten cabbage
Bromine	Bleach, pungent	Methyl methacrylate	Pungent, sulfide like
Sec-Butyl acetate	Fruity	Methyl sulfide	Decayed vegetables
Butyric acid	Sweat, body odor	Naphthalene	Moth balls
Camphor	Medicinal	Nitrobenzene	Bitter almonds
Caprylic acid	Animal like	Phenol	Sweet, tarry odor, carbolic acid
Carbon disulfide	Rotten vegetable	Pinenes	Resinous, woody, pine-like
Chlorine	Irritating, bleach, pungent	Propyl mercaptan	Skunk
Chlorobenzene	Moth balls	Putrescine	Decaying flesh
2-Chloroethanol	Faint, ethereal	Pyridine	Nauseating, burnt
Chloroform	Sweet	Skatole	Excreta, fecal odor
Chlorophenol	Medicinal	Styrene	Penetrating, rubbery, plastic
p-Cresol	Tar-like, pungent	Sulfur dioxide	Pungent, irritating odor
Cyclohexane	Sweetish when pure, pungent when contaminated	Thiocresol	Rancid, skunk-like odor
Cyclohexanol	Camphor, methanol	Toluene	Floral, pungent, moth balls
Cyclohexanone	Acetone-like	Trichloroethylene	Solvent
Diamines	Rotten flesh	Triethylamine	Fishy, pungent
1,1-Dichloroethane	Ether-like	Valeric acid	Sweat, body odor, cheese
1,2-Dichloroethylene	Chloroform-like	Vinyl chloride	Faintly sweet
Diethyl ether	Pungent	Xylene	Aromatic, sweet
Dimethylacetamide	Amine, burnt, oily		

Odor threshold value of common odorants

Compound	mg m^{-3}	ppm	Compound	mg m^{-3}	ppm
Acetic acid	0.043	0.016	2-Hydroxyethyl acetate	0.527	0.114
Acetic anhydride	0.0013	0.00029	Light fuel oil	0.053	Acetone
13.9	4.58	3-	Methylbutanal	0.0016	0.0004
Acrylic acid	0.0013	0.0004	2-Methyl-1-butanol	0.16	0.041
Amyl acetate	0.95	0.163	Methyldithiomethane	0.0011	0.00026
iso Amyl acetate	0.022	0.0038	2-Methyl 5-ethyl pyridine	0.032	0.006
Benzene	32.5	8.65	Methyl methacrylate	0.38	0.085
1,3-Butadiene	1.1	0.455	3-Methoxybutyl acetate	0.044	0.007
1-Butanol	0.09	0.03	1-Methoxypropan-2-ol	0.0122	0.003
2-Butanol	3.3	1	1-Methoxy-2-propylacetate	0.0075	0.0014
2-Butanone (MEK)	0.87	0.27	2-Methyl-l-pentanol	0.096	0.021
Butoxybutane	0.03	0.005	2-Methyl pentaldehyde	0.09	0.02
2-Butoxyethanol	0.0051	0.00097	4-Methyl-2-pentanone (MIBK)	0.54	0.121
2-Butoxyethyl acetate	0.045	0.0063	2-Methyl-2-propanol	71	21.46
Butoxypropanol	0.191	0.0324	α-Methyl styrene	0.021	0.003
Butyl acetate	0.047	0.0066	1-Nitropropane	28.2	7.09
2- (2-Butoxyethocy)ethanol	0.0092	0.0013	1-Octene	0.33	0.066
2,2-butoxyethoxyethyl acetate	0.015	0.0016	2-Octene	0.5	0.1
Carbon tetrachloride	280	40.73	2-Octyne	0.03	0.006
Carbon sulfide	0.0275	0.0102	2,4-Pentanedione	0.045	0.01
m-Cresol	0.0013	0.0003	1-Pentanol	0.02	0.0051
o-Cresol	0.0028	0.0005	Petroleum naphtha	0.2	
p-Cresol	0.0029	0.0006	Phenyl ether	0.0021	0.0003
Cyclohexane	315	83.8	2-Picoline	0.014	0.0034
Cyclohexanone	0.083	0.019	Propanal	0.014	0.0054
Dichloromethane	3.42	0.912	2-Propanol	1.185	0.442
Diesel	0.06		2-Propen-1-ol	1.2	0.47
Dimethyl adipate	7.101	0.913	iso Propylamine	0.158	0.06
Dimethyl glutarate	1.212	0.169	Propylbenzene	0.048	0.009
Dimethyl succinate	0.992	0.152	Propylene-n-butylether	0.206	0.01
1,4-Dioxane	30.6	7.78	Propyl ether	0.024	0.0053
1,3-Dioxolane	56.3	17.02	Styrene	0.16	0.0344
Diphenylmethane	0.41	0.55	1,1,2,2-Tetrachloroethane	1.6	0.21
Ethoxypropanol	0.161	0.035	Toluene	0.644	0.16
Ethoxypropyl acetate	0.0052	0.0008	Trichloroethylene	8	1.36
Ethyl acetate	2.41	0.61	Trimethylamine	0.0026	0.001
Ethyl alcohol	0.28	0.136	Xylene (mixed)	0.078	0.016
2-Ethyl-l-butanol	0.07	0.015	2,3 Xylenol	0.0037	0.0007
2-Ethyl-l-hexanol	0.5	0.086	2,4 Xylenol	0.064	0.0117

(continued)

(continued)

Compound	mg m^{-3}	ppm	Compound	mg m^{-3}	ppm
2-Ethylhexyl acrylate	0.6	0.073			
2-Furaldehyde	0.25	0.058			
1-Hexanol	0.005	0.0011			
Hydrogen sulfide	0.00076	0.0005			

Hedonic score

Description	Hedonic score	Description	Hedonic score	Description	Hedonic score
Cadaverous (dead animal)	−3.75	Stale	−2.04	New rubber	−0.96
Putrid, foul, decayed	−3.74	Fishy	−1.98	Metallic	−0.94
Sewer odor	−3.68	Musty, earthy, moldy	−1.94	Wet paper	−0.94
Cat urine	−3.64	Sooty	−1.69	Medicinal	−0.89
Fecal (like manure)	−3.36	Cleaning fluid	−1.69	Chalky	−0.85
Sickening (vomit)	−3.34	Kerosene	−1.67	Varnish	−0.85
Urine	−3.34	Blood, raw meat	−1.64	Nail polish remover	−0.81
Rancid	−3.15	Chemical	−1.64	Paint	−0.75
Burnt rubber	−3.01	Tar	−1.63	Turpentine (pine oil)	−0.73
Sour milk	−2.91	Disinfectant, carbolic	−1.60	Kippery–smoked fish	−0.69
Stale tobacco smoke	−2.83	Ether, anesthetic	−1.54	Fresh tobacco smoke	−0.66
Fermented (rotten fruit)	−2.76	Burn, smoky	−1.53	Sauerkraut	−0.60
Dirty linen	−2.55	Burnt paper	−1.47	Camphor	−0.55
Sweaty	−2.53	Oily, fatty	−1.41	Cardboard	−0.54
Ammonia	−2.47	Bitter	−1.38	Alcoholic	−0.47
Sulfurous	−2.45	Creosote	−1.35	Crushed weeds	−0.21
Sharp, pungent, acid	−2.34	Sour, vinegar	−1.26	Garlic, onion	−0.17
Household gas	−2.30	Mothballs	−1.25	Rope	−0.16
Wet wool, wet dog	−2.28	Gasoline, solvent	−1.16	Beery	−0.14
Mouse-like	−2.20	Animal	−1.13	Burnt candle	−0.08
Burnt milk	−2.19	Seminal, sperm-like	−1.04	Yeasty	−0.07
				Dry, powdery	−0.07
Cork	0.19	Crushed grass	1.34	Maple syrup	2.26

(continued)

(continued)

Description	Hedonic score	Description	Hedonic score	Description	Hedonic score
Black pepper	0.19	Celery	1.36	Pear	2.26
Musky	0.21	Green pepper	1.39	Caramel	2.32
Raw potato	0.26	Tea leaves	1.40	Coffee	2.33
Eggy (fresh eggs)	0.45	Aromatic	1.41	Meaty (cooked, good)	2.34
Mushroom	0.52	Raisins	1.56	Melon	2.41
Beany	0.54	Cooked vegetables	1.58	Popcorn	2.47
Geranium leaves	0.57	Clove	1.67	Minty, peppermint	2.50
Grainy (as grain)	0.63	Nutty	1.92	Lemon	2.50
Dill	0.87	Coconut	1.93	Fragrant	2.52
Woody, resinous	0.94	Grapefruit	1.95	Fried chicken	2.53
Soapy	0.96	Perfumery	1.96	Cinnamon	2.54
Laurel leaves	0.97	Peanut butter	1.99	Cherry	2.55
Eucalyptus	0.99	Spicy	1.99	Vanilla	2.57
Molasses	1.00	Banana	2.00	Pineapple	2.59
Incense	1.01	Almond	2.01	Apple	2.61
Malty	1.05	Sweet	2.03	Peach	2.67
Caraway	1.06	Buttery, fresh butter	2.04	Violets	2.68
Soupy	1.13	Grape juice	2.07	Fruity, citrus	2.72
Bark, birch bark	1.18	Honey	2.08	Chocolate	2.78
Anise (liquorice)	1.21	Cedarwood	2.11	Floral	2.79
Oak wood, cognac	1.23	Herbal, green, cut grass	2.14	Orange	2.86
Seasoning (for meat)	1.27	Cologne	2.16	Strawberry	2.93
Leather	1.30	Fresh green vegetables	2.19	Rose	3.08
Raw cucumber	1.30	Fruity, other than citrus	2.23	Bakery (fresh bread)	3.53
Hay	1.3.1	Lavender	2.2.5		

Appendix II: Odor Wheel [27]

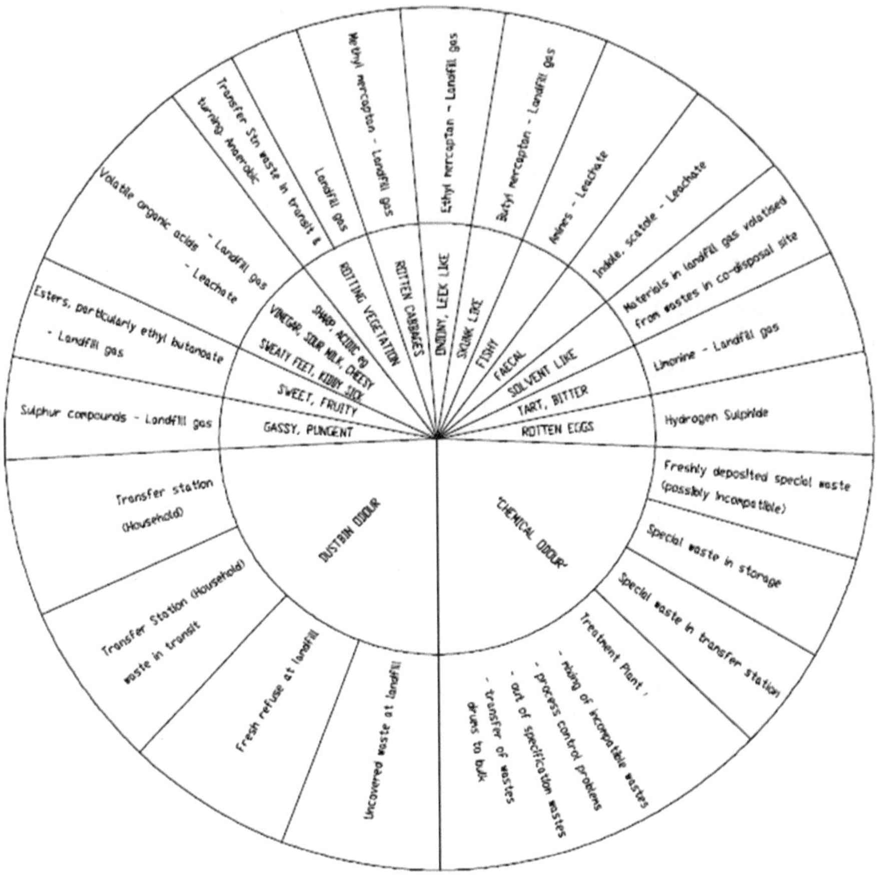

References

1. http://www.extension.iastate.edu/airquality
2. K. Bauer, D. Garbe, H. Surburg, "Common fragrance and flavor materials", VCH Verlagsgesselschaft, FRG, 1990
3. EPA., Odour impacts and odour emission control measures for intensive agriculture Final Report. R&D Report Series No. 14, European Community European Regional Development Fund and EPA (Environmental Protection Agency) (2001a)
4. A.Yuwono, P. Schulze Lammers, "Odour Pollution in the Environment and the Detection Instrumentation". Agric. Eng. Int.: CIGR J. Sci. Res.Dev. (2004)
5. J.W. Gardner, P.N. Bartlett, *Electronic noses principles and applications* (Oxford University Press, Oxford, 1999)

6. ASTM International, E1593-94: Standard practice for assessing the efficiency of air freshener products in reducing sensorily perceived indoor air malodour intensity. Philadelphia, USA, 1999
7. Performance verification of air freshener products and other odour control devices for indoor air quality malodours, The 8th Workshop on Odour and Emissions of Plastic Materials, Kassel, Germany (2006)
8. C.M. McGinley, "Odour testing bio-solids for decision making', Water Environment Federation Specialty Conference: Residuals and Bio-solids Management Conference Austin, TX, 3–6 March, 2002
9. C.M. McGinley, T.D. Mahin, R.J.Pope, Elements of successful odour/odour laws. Water environment federation odours and voc emissions 2000 specialty conference, Cincinnati, 2000
10. H.R. Moskowitz, A. Dravnieks, C. Gerbers, Odour intensity and pleasantness of butanol. J. Exp. Psychol. **103**, 216–223 (1974)
11. M. Meilgaard, G.V. Civille, B.T. Carr, *Sensory evaluation techniques* (CRC Press, Boca-Raton, 1987)
12. B.G. Green, G.S. Shaffer, M.M. Gilmore, Derivation and evaluation of a semantic scale of oral sensation magnitude with apparent ratio properties. Chem. Senses **18**, 683–702 (1993)
13. "Odour", ASHRAE Fundamentals Handbook (SI), Ashrae Handbook (2001)
14. A. Dravnieks, Evaluation of human body odours, methods and interpretations. J. Soc. Cosmet. Chem. **26**, 551 (1975)
15. INFOMETRIX. APPLICATION NOTES, "Electronic nose instrumentation" (1999), www.infometrix.com
16. Elements of successful odour/odour laws by Charles M. McGinley, Thomas D. Mahin and Richard J. Pope. WEF Odour/VOC 2000 specialty conference Cincinnati, OH, pp. 16–19, 2000
17. M. Susan Brewer, K. R. Cadwallader, "Overview of odor measurement techniques", Department of Food Science and Human Nutrition University of Illinois 1302 W. Pennsylvania Ave., Urbana, IL 61801, 2003
18. Sven Nimmermark, "Odour release, dispersion and influence on human well-being with specific focus on animal production", Department of Agricultural Bio-systems and Technology, Doctoral thesis, Swedish University of Agricultural Sciences, Alnarp 2004
19. Japan association of odour environment. http://www.orea.or.jp
20. Guideline on Odour Pollution and Control, central pollution control board, Govt. of India, Delhi, 2008
21. M. Osako, "Prospects for Development of Simplified Evaluation Methods used in Odour Management". National Institute for Environmental Studies 16–2 Onogawa, Tsukuba, Ibaraki Pref. 305–8506, Japan
22. "Quality control of olfactometry at SRI and in Europe**"**, Robert Sneath, Silsoe Research Institute, Wrest Park, Silsoe, Bedfordshire, MK43 9JY
23. Odour methodology guideline, Department of Environment Protection, Perth, Australia, (2002)
24. "Odour guidance 2010", SEPA: Scottish environment protection agency, Scotland, 2010
25. A. Dravnieks, T. Masurat, R.A. Lamm, Hedonics of odours and odour descriptors. J. Air Pollut. Control Assoc. **34**(7), 752–755 (1984)
26. Y. Nagata, N. Takeuchi, Measurement of odour threshold by triangle odour bag method. Bull. Jpn. Environ. Sanitation Center **17**, 77–89 (1990)
27. Odour guidance for waste sites, version 3.0, July 2002, Internal guidance for the regulation of odour at waste management facilities (2002)

Chapter 5
Toward Sensor to Reproduce Human Sense

5.1 Introduction

We are constantly engaging in a chemical communication with environment. Our bodies smell and by doing so we release relevant information about ourselves. Some of this information is released in the form of volatiles and some is even detectable by the human nose. Other signals are only perceivable by biological antennae. Bacteria communicate using chemical signals, by releasing and receiving signaling molecules in what is known as quorum sensing. They don't just communicate amongst themselves but also interact with signals sent by their human host.

The human nose involved in the four functions of odor sensing: detection, recording, memory search and recognition, which are replicated by machines, identify as electronic noses (E-noses). This e-nose which contains chemical sensors, which is used for detecting and recording; and in addition artificial intelligence (AI) software artificial neural network (ANN) is used for memory search and identification [1].

Smells reach the olfactory sensory neurons by way of two pathways. The first pathway is through nostrils. The second pathway is through a channel that connects the roof of the throat region to the nose. When human chew food, aromas are released that access the olfactory sensory neurons through this channel. If the channel is blocked, such as when noses are stuffed up from a cold, odors cannot reach the sensory cells and human ability to enjoy a food's flavor is lost. This is the technique in which, senses of smell and taste work closely together. Without the olfactory sensory neurons, familiar flavors like pineapple or strawberry would be hard to differentiate. People who go to the doctor because they think they've lost their sense of taste are surprised to learn that they have a smell disorder instead.

Our sense of smell is also influenced by something called the common chemical sense. This sense involves thousands of nerve endings, especially on the moist surfaces of the eyes, nose, mouth, and throat. These nerve endings help us sense irritating substances such as the tear-inducing power of an onion or the refreshing cool of peppermint.

Some of the fundamental signal-analysis techniques that find wide application with bio-signals are discussed.

H. K. Patel, *The Electronic Nose: Artificial Olfaction Technology*,
Biological and Medical Physics, Biomedical Engineering,
DOI: 10.1007/978-81-322-1548-6_5, © Springer India 2014

5.2 Electronic Nose

During the last decades artificial nose technology has provided many successful examples of industrial applications. As this technology becomes cheaper and more accessible, the possibilities of potential use in medical and clinical studies have also increased.

When chemicals are detected by particular cells in the nose resulting in nerve impulses which are sent to the brain for interpretation is called recognition of smell. Smell is responsible for the perception of 75–80 % of all we taste. Smell results from molecule of various chemicals floating through air. All substance does not have smell, only few which include chemicals that are volatile in nature.

E-nose was developed with characteristic alike an inexpensive, quick, and portable device capable of quantifying complex mixtures of volatile compounds. The realization of the E-nose was made possible by using the human olfactory system as a paradigm.

There are several sensors which are specific to particular compounds or classes of compounds, but using compound-specific sensors to monitor changes in breathing air is a task which would involve the use of several hundreds of sensors. In addition, compound-specific sensors are subject to interference from molecules of structure similar to the target molecule. Analytical instruments such as gas chromatography—mass spectrometry (GC–MS) have multiple capabilities, but are not practical for continuous monitoring. Work on the E-nose was designed to fill the gap between individual, chemically specific devices and analytical instruments such as GC–MS which have multi-compound capability.

The main section of an E-nose is an array of non-specific chemical sensors. Sensor array output stimulates by an odor analyte and a pattern of characteristic response is generated. E-nose sensor made of a compilation of technologies, also note that in every case a specific physical property is measured and a set of signals is generated. The last stage is pattern recognition (PARC) process and it's also similar to biological olfaction, where a sensor type responds to more than one odorant and one odorant type activates more than one sensor. So this set of process collectively, activated sensors and their signals characterize the odor (sometimes referred as an odor fingerprint). Therefore, an important difference between E-noses and analyte detectors such as gas chromatographs, is that whereas the latter are aimed at identifying the components that contribute to an odor, E-noses can be used to identify, as a whole, the mixture of components that together form an odor [2].

Multi-compound sensing may be accomplished using an array of partially specific sensors, or an E-nose. In an E-nose, the distributed response of an array of several sensor heads is used to identify the changes in the composition of a gaseous environment. This is accomplished by comparing the response of the array to a baseline response of known, "good" air. Sensors are not particular to any one gas; it is in the use of an array of sensors with a different sensing medium on each that gases and gas mixtures can be recognized by the pattern of response of the array.

So far, sensor arrays have been used only in fairly restricted applications, such as the controlled atmosphere of quality control for beer or coffee manufacture. There are commercially available E-noses, which are neither miniature nor low power.

Electronic noses are designed to mimic the human sense of smell by providing an analysis of individual chemicals or chemical mixtures. They offer an efficient way of analyzing and comparing odors. Electronic noses have yet to reach the capability of decomposing odors into their chemical components. E-noses primarily detect and discriminate between odorants they previously "learned".

In the majority of E-nose applications there is a set of desired properties that tend to be common.

Rapid response—The sensors should be able to react to and recover from an exposed odor within an acceptable time frame. This is especially important in applications that integrate E-nose with a robotic system, such as a mobile robot that should move around in an environment and measure odor gradient.

Low power consumption—In most realistic systems there is a limit of power, and thus the power consumption of the sensors should be relatively low. The headspace, containing the sensor array is likely to involve other electrical equipment such as pumps and valves which often share the same power supply.

Compact size—Smaller sensor size facilitates the integration of sensors in a variety of platforms, including portable E-noses.

High sensitivity—The sensors should exhibit a high sensitivity to different odorants and different concentrations of the same odorant.

Reliability—Gas sensors should behave as expected particularly over long periods of time.

Robustness—Unwanted effects, from humidity and physical motion, should not disturb the results from the sensor readings.

Electronic noses basically try to mimic the principle components of a mammalian nose. The fundamental processes that must occur for a mammalian nose to detect and identify an odor can be summarized as follows:

1. Sniffing
2. reception and binding
3. stimulus
4. transmission
5. recognition
6. action
7. cleansing

The parallel mechanisms in an E-nose would be the following:

1. drawing in some gas from the i.e. environment, sample of food etc.,
2. confer the vapor molecules to react with an array of sensors,
3. detecting signals indicating the reaction of the sensors to the vapor,
4. transmitting these signals to a neural network or some other PARC mechanism,

5. matching the signal pattern with signatures of known chemical vapors and thus identifying the chemical composition of the vapor,
6. taking some form of action based on the identification, and
7. cleaning the sensor array so that recognition process can take place again [3].

Sensing system and the automated PARC system are two main components of an E-nose. Human nerve olfactory performs functions, very similar functions performed by the sensor array in E-nose. So, the sensor array may be considered the heart and most important component of the E-nose. The sensing system can be an array of special sensing elements (e.g., gas sensors), where each element measures a different property of the sensed odor or, the sensors respond to a complex odor with overlapped sensitivity, which results in the sensor array producing a distinguishing pattern. The PARC systems include the feature extraction step, which extracts useful information from the sensor responses and the PARC algorithm. The PARC system is completed by interfacing with the computer central processing unit (CPU), identification library and detection software that serve as the brain to process input data from the sensor array for successive data analysis.

The sampling system is also important though it is listed as third necessary part of an E-nose. The sampling system main lookout, that samples are supplied to the sensor array in a reproducible way, and conditioned if necessary to adjust concentration, temperature, water vapor concentration, etc. Many sampling systems also give automated measurement of a series of samples, though this is not a major requirement. A sampling system should deliver a vapor sample to a sensor array in a reproducible way. Its purpose is to reduce sample-to-sample variation that may result from differences in humidity, temperature, concentration, etc., as well as to preprocess the sample in any way that increases the quality of output data. Samples often contain substances that are common to all, and, although the sensors are dominated by them, they do not contribute to inequity. Numbers of criteria should be accomplished by a good sensor. Main and important is the sensor should have highest sensitivity to the target group of chemical compound(s) intended for detection and with a threshold of detection similar to that of the human nose.

Electronic nose sensor operation involves interactions between gaseous molecules and sensor-coating materials which modulate electrical current passing through the sensor, detectable by a transducer that converts the modulation into a recordable electronic signal [4]. Different number of electrochemical sensors (e.g., metal-oxide gas sensors, metal-oxide semiconductor field effect transistors, conducting polymer gas sensors, acoustic wave gas sensors, quartz crystal microbalance sensors, surface acoustic wave devices, field-effect gas sensors, electrochemical gas sensors, pellistors, fiber-optic gas sensors) and many different types of sensor-coating materials are classified according to additive doping materials, the type and nature of the chemical interactions, the reversibility of the chemical reactions and running temperature.

Various types of transducer devices in electronic-nose sensors are categorized according to the nature of the physical signal they measure. Transduction principles based on electrical measurements, including changes in current, voltage, resistance or impedance, electrical fields and oscillation frequency are most common methods which are utilized. Other methods are measurements of mass changes, temperature changes or heat generation. Optical sensors measure the modulation of light properties or characteristics such as changes in light absorbance, polarization, fluorescence, optical layer thickness, color or wavelength (colorimetric) and other optical properties [5].

Note that the effect of slight changes in the specific amounts of chemical type within an odor mixture often can be detected by the human nose as a change in odor by trained panel experts, but changes in odorless materials are not detectable. The added advantage of E-nose is detecting definite odorless compounds that are not detectable by the human nose. Each vapor presented to the sensing system generates a signature or "fingerprint". Presenting many different chemicals to the sensor yields a database of fingerprints, which the pattern-recognition system uses to recognize and automatically identify each chemical.

Many E-noses are commercially offered today and have an extensive range of applications in different markets and industries ranging from food processing, industrial manufacturing, quality control, environmental protection, security, safety and military applications to various pharmaceutical, medical, microbiological and diagnostic applications.

Electronic noses with different types of sensor arrays are differentially responsive to a wide variety of possible analytes and have a number of advantages over classical analytical instruments. E-nose sensors do not require chemical reagents, have high-quality sensitivity and definite, give rapid results, and allow non-destructive sampling of odorants or analytes e-noses generally are far less expensive than analytical systems, easier and cheaper to operate, and have greater potential for portability and field use compared with complex analytical laboratory instruments. E-noses have good potential to be used in the long term by inexperienced users for numerous practical applications in residential and public settings. Several disadvantages of e-nose sensing contain problems with reproducibility, improvement, effects of humidity and temperature on the sensor responses, and incapability to recognize individual chemical class with sample gases. Thus, E-noses may never completely replace complex analytical equipment or odor panels for all applications, but offer quick real-time detection and discrimination solutions for applications requiring accurate, rapid and repeated determinations.

The E-nose consists of five key components: sampling chamber, sensor chamber, data acquisition system and controller unit, power supply and graphic user interface on a computer. Figure 5.1 shows the block diagram of the E-nose system.

Human nose is elegant, sensitive, and self-repairing, but the E-nose sensors do not fatigue or get the "flu". Further, the E-nose can be sent to detect toxic and or else hazardous situations that humans may desire to keep away from i.e., sensors can detect toxic carbon monoxide, which is odorless to humans.

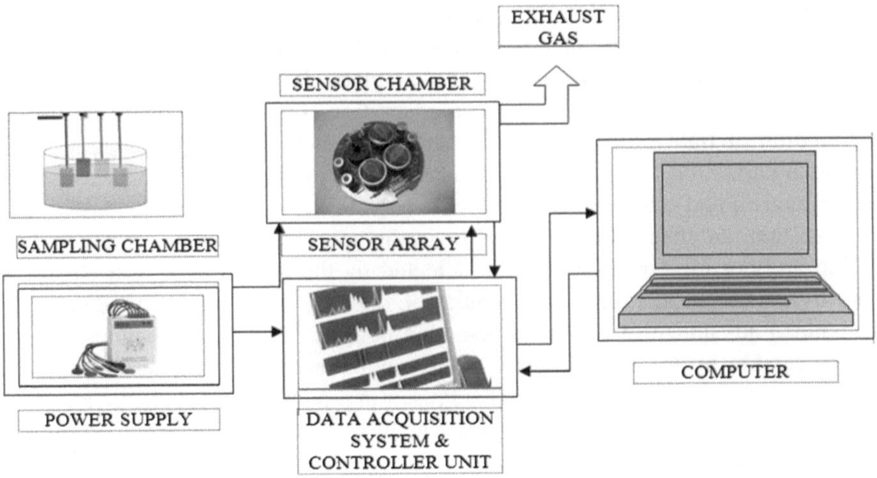

Fig. 5.1 Block diagram of E-nose system

5.3 Electronic Tongue

Taste plays an essential role in food selection and consequently overall nutrition. Our sense of taste helps us to gain information to form a picture of the world by sampling chemicals from our environment. Electronic tongue (E-tongue) is an analytical device that is used to analyze liquid samples. The principle is based on the organizational principles of biological sensory systems.

The E-tongue mirrors the three levels of biological taste recognition: the receptor level (taste buds in humans, probe membranes in the E-tongue); the circuit level (neural transmission in humans, transducer in the E-tongue) and the perceptual level (cognition in the thalamus in humans, computer and statistical analysis in the E-tongue.

Electronic tongue is an analytical instrument consists of an array of non-specific, selective, chemical sensors with partly specificity to different components in solution, and an appropriate method of PARC and/or multivariate calibration for data processing. Properly configured and calibrated E-tongue is capable of recognizing the quantitative and qualitative composition of multi-component solutions of different natures. The E-tongue is a set of potentiometric chemical sensors, applicable for liquid analysis. Sensor arrays include different types of sensors, conventional ones, specially designed non-specific sensors with enhanced cross-sensitivities or classical electrochemical electrodes are used depending on the task, sensor stability or cross sensitivity. The second essential part of the E-tongue is data processing. Since the number of sensors in the array of an E-tongue can reach 40, each of them producing a complex response in the multi-component environment, a relevant multidimensional data processing must be performed.

Taste sensor or E-tongue is a logical tool including an array of non-specific, low selective chemical sensors with cross-sensitivity to different components in solution accompanied by an appropriate method of PARC and/or multivariate calibration for the data-processing. The stability of sensor behavior and enhanced cross-sensitivity is a critical criterion, which is understood as reproducible response of a sensor to as many species in solution as possible. If properly configured and trained the E-tongue has the potential to determine quantitative composition and to recognize complex liquids of different nature. The sense of taste may have two meanings. One aspect devotes to the five basic tastes of the tongue; sour, salt, bitter, sweet, and 'umami'. These tastes are sensed from different, discrete regions on the tongue including specific receptors known papillae. The other aspect denotes the perception obtained when food enters the mouth [6].

When chemicals are detected by specific cells on the tongue resulting in nerve impulses which are sent to the brain for analysis than taste occurs. Taste reception happens at the apical tip of taste cells that form taste buds. Each onion shaped taste bud is composed of 50–100 taste cells that have microvilli. Each single taste bud contains 50–100 taste cells signifying all 5 taste sensations. Implanted in the cell membranes of these taste cells are receptor proteins. Each of the taste receptors are transmembrane proteins which role either by physically binding to a flavor ingredient (sweet, bitter and umami) or by acting as a channel to allow ions to flow directly into a taste cell (salty and sour). This interaction triggers a signaling cascade that culminates with signals to the brain through a network of taste nerve fibers.

Gustation, the sense of taste, is important for the location and identification of food and can be a very highly developed sense. e.g., some Cyprinids have sensitivity to sugars and salts that are 512 and 184 times higher than humans, respectively. Different types of olfactory receptors, taste receptors can be found in more than one location on the body. Besides being found in the mouth, pharynx, gill arches and skin, some fish, such as the Siluri formes have well developed barbells that have dense concentrations of taste receptor cells. Figure 5.2 shows the taste bud cell.

The performance of an artificial sense such as the E-tongue can be considerably enhanced by the combination of sensors based on different technologies. The reason is, of course, that for each new measurement principal added, also a new dimension of information is added. A natural extension of this fundamental concept is the combination of different artificial senses. This is especially important when estimating the quality of food, since the guide is the impression of the human being using all five senses [6].

A taste sensor system is used to categorize the different basic taste sensations (sour, salt, bitter, sweet and "umami"), and the results are compared with human test panels. An E-tongue classifies a quality of one or another kind in food, drinks, water, process fluids etc., and the results are not necessarily compared with human sensations, but with other quality properties of the sample [1]. Figure 5.3 represents the schematic analysis of the E-tongue system.

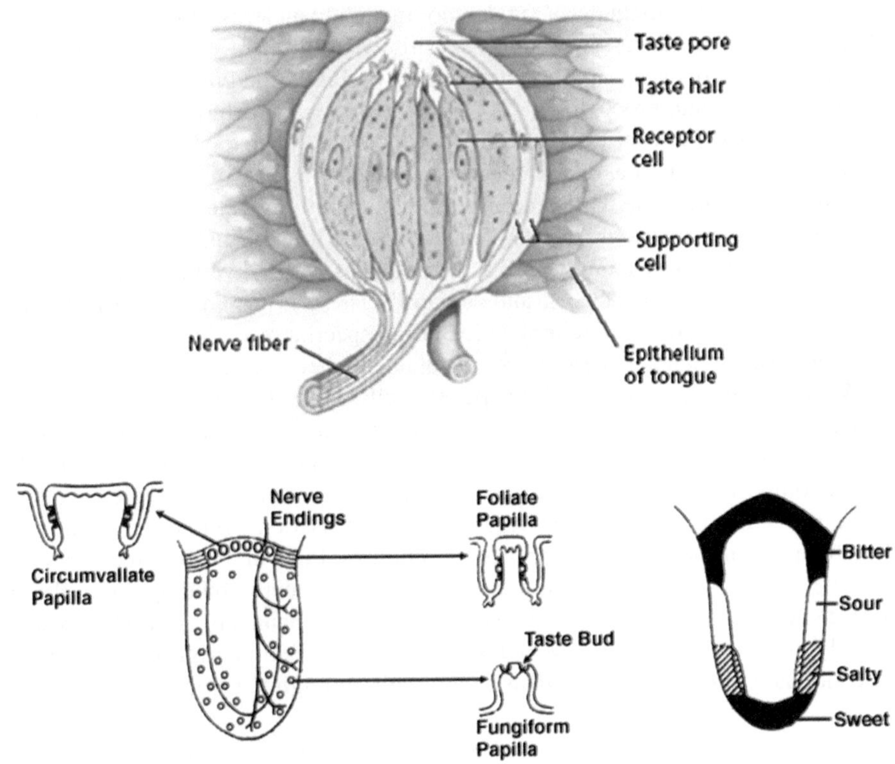

Fig. 5.2 Taste bud cell

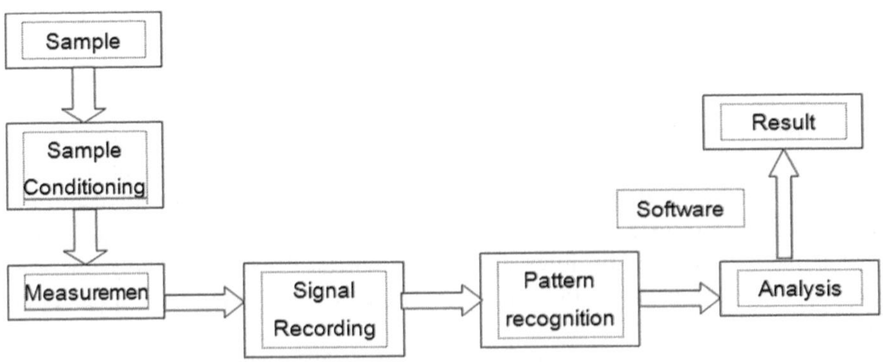

Fig. 5.3 E-tongue analysis schematic

Electronic tongue has found wide applications in different areas including foodstuff, industrial, clinical and environmental analysis. In foodstuff, the E-tongue is used for quality control of processing and storage of mineral water, wine, coffee, milk, juice and tea. It is also used for the recognition, identification, classification and discrimination of different kinds of liquid foodstuff and beverages.

E-tongue or taste sensors are committed to the automatic analysis of complicated composition samples, to the recognition of their characteristic properties, and generally they are assigned to fast qualitative analysis. The fundamental rule used in E-tongues is based on electrochemical methods such as Potentiometry [7, 8] and voltammetry [9]. Also E-tongue has been developed with the principles based on optical methods and impedance to mimic the basic taste sensations experienced by humans [10, 11].

Potentiometry is based on zero current technique, in which a potential across a surface region on the working electrode is measured [12]. The various types of membrane materials have been developed with different detection properties. A Potentiometry devices are widely used for the measurement of a large number of ionic classes and in that important is the pH-electrode; other examples are electrodes for calcium, potassium, sodium and chloride.

The E-tongue tools offer the potential to provide taste evaluation during formulation development or to be used as a quality-control tool, and thus significantly reduce the amount of human testing. Another important benefit includes the reduction of the dependence on human panels. Human panels in general present several difficulties including health concerns, fatigue of tasters, maintaining the motivation for tasting unpleasant compounds and the lack of analytical standardization. While it is necessary to develop good tasting treatments, the use of sensory panelists is very difficult in this industry. This is due to the potential toxicity of drugs and subjectivity of taste panelists. Problems in recruiting taste panelists, enthusiasm and panel maintenance are significantly more difficult when working with unlikable products.

5.4 Key Benefits of E-tongue Taste Evaluation

1. Helping to quantify bitterness of preparation actives when limited basic taste information is available, especially if the supply is limited.
2. Developing suitable matching bitter placebos for blinded clinical testing.
3. Developing optimized taste-masked formulations.
4. Measuring efficiency of complication/coating within formulation.
5. Conduction comparator studies (benchmark analysis).
6. Serving a quality control traction for flavored product and excipient.

Difference is found in sense receptors, as in taste buds receptors on the tongue, and smell receptors, embedded high in the lining of the nasal cavity. Gustatory receptors respond to direct contact with water-soluble materials (e.g., sugar) and olfactory receptors to generally water-insoluble, vaporous materials that may even arise from a distant source. Many establishments choose to regard smell as distance chemoreception and taste as contact chemoreception.

5.5 Applications

- Biomedical and biotechnology applications: Urine analysis, Herbal analysis, drug analysis.
- Food analysis: wine evaluation for taste and flavor, Alcohols in beverages, freshness of milk, coffee brand analysis; analyze enhanced flavored level i.e., in flavored milk, fruit juice.
- Pharmaceuticals industries: taste quantifying tablets, syrups, powders, capsules.
- Environment monitoring
- Safety
- Chemical industry

5.6 Machine Olfaction

Smell is still anonymous to scientists in somehow, which cannot be studied with straightforwardness in vertebrates. Another problem is that the sense of smell is poorly developed in human beings in comparison with the same in many vertebrates [13]. This makes realization of an artificial olfactory system a challenging task. An artificial olfactory system (commonly known as E-nose) provides a low cost alternative to identification, quantification and characterization of odors.

The inspiration behind the mechanisms of machine olfaction is human olfaction. So to understand the operation of an E-nose it is important to first understand the biological act of smelling. With the human olfactory system that enables us to characterize an odor. Therefore, a description of the human olfactory system and the properties of odorous molecules are discussed in Chaps. 3 and 4 before discussing the so-called method of machine olfaction.

The Artificial nose (Machine olfaction) is a system consisting of three functional components that operate in sequence on an odorant sample- a sample handler, an array of gas sensors, and a signal-processing system. Odorant uniqueness is the output of the E-nose and an estimate of the concentration of the odorant, or the characteristic properties of the odor as might be perceived by a human. The olfactory system detects small differences in the composition of natural odorants, made up of hundreds of molecules. Odorous quality is theoretically represented by a combinatorial code: activation of distinct but overlapping subsets of olfactory receptors resulting in activation of a distinct subset of glomeruli in the main olfactory bulb.

Fundamental to the artificial nose is the idea that each sensor in the array has different sensitivity. For example, odorant No. 1 may produce a high response in one sensor and lower responses in others, whereas odorant No. 2 might produce high readings for sensors other than the one that "took" to odorant No. 1. Human olfaction and machine olfaction have different mechanisms to sense smell, and thus the quantitative description of an odorant can also be very different for human

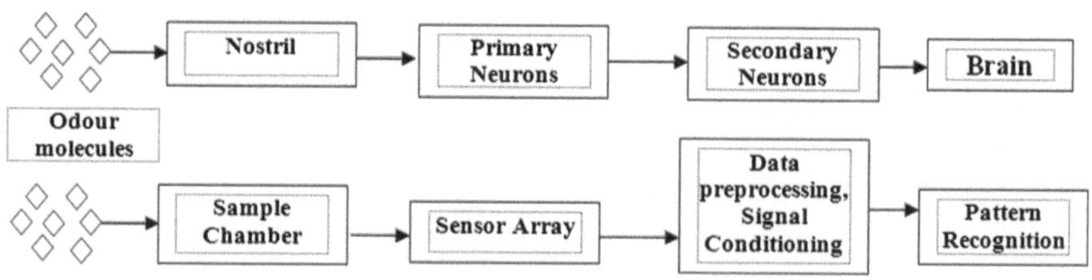

Fig. 5.4 Comparison of olfactory system with the artificial olfaction system

and artificial noses. When artificial noses are used to characterize or compare human smell performance, the signals obtained with these devices need to be calibrated and matched to what is considered a normal perception by a human. What's important is that the pattern of response across the sensors is separate for different odorants. This distinguished ability allows the system to identify an unknown odor from the pattern of sensor responses. Each sensor in the array has a unique response profile to the spectrum of odorants under test. To identify the odor all the sensors in the array pattern response is used. While talking about Machine olfaction, there is a comparison of biological olfaction can be found required and useful, in particular the human olfactory system and the respective cognitive processes. It's shown in Fig. 5.4.

There are significant similarity between machine olfaction system and the "Human-nose" constructed by Nature. Sniffing, this moves air samples which include molecules of odors to the thin mucus layer lining the olfactory epithelium in the upper portion of the nasal cavity. The molecules of odor interact with the membrane bound receptor proteins of the olfactory cells. The human nose uses the lungs to bring the odor to the epithelium layer; in the machine olfaction system uses Bio VOC/Breath exhaled pump.

In the olfactory system, when odor molecules communicate with the different proteins, a series of nerve impulses entitle electrical stimulus generated by the olfactory neurons, feed into the olfactory bulb. The overall function aim is to reduce the noise by compressing the signals and amplifying the output. In short, the human nose has mucous, hairs, and membranes to work as filters and concentrators, while in the machine olfaction system an inlet sampling system that provides sample filtration and conditioning to protect the sensors and develop selectivity.

The human epithelium contains the olfactory epithelium, which contains millions of sensing cells, selected from different genotypes that interact with the odorous molecules in unique ways. The machine olfaction system contains various types of sensors that interact differently with the sample. The chemical responses are converted in to the electronic nerve impulses by human receptors. The unique patterns of nerve impulses are propagating by neurons through a complex network before reaching the higher brain for interpretation. Similarly, the chemical sensors in the machine olfaction react with the sample and produce electrical signals.

Table 5.1 Comparison between human and artificial nose (E-nose) [14]

Human	Electronic
10 million receptors, self generated	5–100 chemical sensors manually replaced
10–100 selectivity classes	5–100 selectivity patterns
Initial reduction of number of signals (1,000–1)	"Smart" sensor arrays can be mimic
Adaptive	Perhaps possible
Saturates	Persistent
Signal treatment in real time	Pattern recognition hardware may do this
Identifies a large number of odors	Has to be trained for each application
Cannot detect some simple molecules	Can detect also simple molecules (H_2, H_2O, CO_2...)
Detects some specific molecules	Not possible in general at very low Concentrations
Associative with sound, vision, experience etc.	Multisensor systems possible
Can get "infected"	Can get poisoned

A computer reads the unique pattern of signals, and interprets them with some form of intelligent pattern classification algorithm. Table 5.1 gives comparison between the human olfaction system and electronic olfaction system [14].

The general performance of E-noses are yet comparable to that of a human nose, already they can detect substances that are "odourless" to a human such as carbon monoxide. E-noses are also capable of better detecting varying levels of concentrations of odors.

Electronic nose research was inspired by the mechanisms involved in human olfaction. Our sense of smell is able to recognize and discriminate extraneous volatile compounds of diverse molecular structure with high sensitivity and accuracy. Mimicking the perception of odors by humans is the ultimate challenge for machine olfaction and an extremely complex regression problem.

5.7 From Reception of Odor Substances to Perception

Sensations take place any time when a stimulus activates one of the receptors. Perception occurs when we apply our experience to understand sensations. A detail of sensor stimuli comes from millions of sensory receptors in eyes, ears, nose, tongue, skin, muscles, joints, and muscles. Different receptors detect different types of physical energy, such as light waves, mechanical energy, chemical energy, and heat energy. Receptors transduce energy from one form into another. *Transduction* is defined as the transformation of stimulus energy to the electrochemical energy of neural impulses. Excluding for impulses of olfaction/smell transmitted directly to the olfactory bulbs on the underside of the cortex, impulses from sense organs are transmitted to the thalamus before the cortex. The cerebral

Fig. 5.5 The path of sensory information

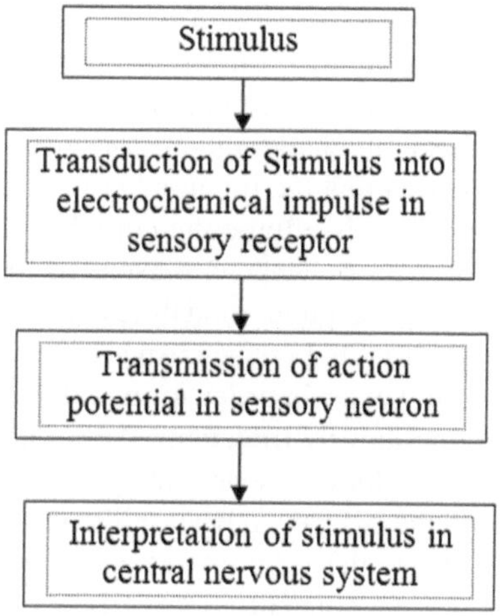

cortex puts all the sensory information together and acts on it. Different areas of the cortex translate neural impulses into different psychological experiences, such as odor or touch. *Perception* is the process of selecting, organizing, and interpreting sensations, enabling to recognize meaningful objects and events. Perception results from the interaction of many neuron systems, each performing a simple task. A sensation occurs any time a stimulus activates one of the receptors. Perception occurs when we apply our experience to interpret sensations. Perception is composed of sensations to which the brain reacts. But there is no clean separation of sensations and perceptions. The precise quantification of olfactory perception is an essential step for the evaluation, diagnosis, and treatment of smell disorders and contributes to the success of a corresponding therapy. Figure 5.5 represented the path of sensory information.

The essential way in which smell differs from the other special senses are deficient in terms of an efficient method of classifying and analyzing different types of smells. To some extent this is because human ordinarily pay little conscious attention to smell, having enough to do in coping with more interesting information pouring in from our eyes and ears. The most intimately connected of all our senses with the brain is the sense of smell. Although, the sense of smell is very complex and its function are also. The method by which the odor receptor cells interact with odor-causing molecules is still unknown, but studies of odors and the structure of the odor-causing molecules have revealed some correlations. The nose is generally ignorant without special instruction whether an exacting smell is pure, in the sense that only one type of molecule is there, or a mixture. Many natural odors that look like perfectly unique and pure, like that of pineapple, are in fact composed of dozens of molecules, several of which taken by themselves are rather unpleasant, and cannot be detected for what they are in the whole group. Tables 5.2 summarize the human senses and their sensory information.

Sensory perception relies on the way sensory signals are transformed by neural circuits in the central nervous system. The relation between a given stimulus parameter and perception were always one-to-one; it would be easy to predict the effects of changing one or more stimulus parameters. However, studies have revealed that the relation between the parameters of a sensory stimulus and what is perceived is often quite complex. In other words, what we perceive is not a direct reflection of what is present in the environment. Olfaction is different from the other perceptual modalities in ways that have serious import for the study of cognition and consciousness.

However, in order for an odorant to be smelled, it must first reach the olfactory epithelium by traversing the nasal passage and the Mucosa layer within our nostrils. Odorants reach the olfactory epithelium and olfactory receptor neurons orthonasally or retronasally. Orthonasally (from the front of the nose via the nostrils) the odorant reaches the epithelium either through diffusion from high levels of concentration to lower levels or through actively sniffing the odorant. Alternatively, an odorant might arrive from the back of the throat via retro-nasal olfaction [15].

Sensory receptors convey the energy of stimuli into membrane potentials and transmit signals to the nervous system. Sensory receptors perform four functions in this process: sensory transduction, amplification, transmission, and integration. Humans have several types of sensory nerves: photoreceptors (for light), mechanoreceptors (for touch, sound and equilibrium), chemoreceptors (for smell and taste), thermoreceptors (for heat), and nociceptors (for pain). Classification of sensory systems is by type of stimulus. Each sense cell is responsive to one sort of energy change and can cause only one sensation, though the intensity of sensation depends on the nerve's threshold value. Normal sensing happens by stimuli interacting with a biological system, which elicits a positive or negative response. The receptors that are essential to an organism understanding its environment and surrounding and of most interest to the engineering community for mimicry are grouped into the class known as extroreceptors. The three classes of receptors are:

Table 5.2 Human senses and their sensory information

Olfaction	Source of sensory information	Molecules of the substance being sensed
	Receptor organ	The nose
	Receptor cells	Receptors in each nostril can sense different substances on the basis of molecular shapes
Taste	Source of sensory information	Molecules of the substance being sensed
	Receptor organ	Taste buds on the tongue (additional receptors in the mouth and throat)
	Receptor cells	Taste receptor are sensitive to five basic taste: Sweet, sour, salty, bitter and umami
Skin senses	Source of sensory information	Touch, pressure, warmth, cold and pain
	Receptor organ	The skin
	Receptor cells	Receptors that code for touch, pressure, warmth, cold and pain

Extroreceptors–external–Chemoreceptors, Electroreceptors, Magnetoreceptors, mechanorecptors, Photoreceptors, Thermoreceptors

Proprioceptors—internal–vestibular, muscular etc.

Interoreceptors—internal without conscious perception-Interoreceptors detect stimuli originating inside the body blood pressure, oxygen tension etc.

Table 5.3 summarizes the receptor family and their detection of sensations. Different forms of energy detected by sensory receptors are beginning of the sensations. This energy is transformed to action potentials that travel to suitable regions of the brain. Once the brain is sensitive of sensations, it interprets them, giving the perception of stimuli. Perceptions such as colors, smells, sounds, and tastes are creation formed in the brain and do not exist outside of it. We do perceive objects using our sense of smell, but the objects of these experiences are not the same as when we sense objects through touch, sight, or audition. The objects of vision have sharply delineated spatial characteristics and the objects of audition have clear temporal boundaries. Smells, however, seem to be neither spatially nor temporally bound objects. The object of olfactory perception is unlike our common sense conception of ordinary three-dimensional objects, identified visually. Smells are not ordinary objects, but the chemical structure of simple molecules or mixtures within odor plumes. Our olfactory perceptions are of the material objects of chemistry. We smell the chemical structure of matter. Chemicals compounds, which could lead to smells, constantly surround us. Nonetheless of us are able to separate out different smells from the environment.

The olfactory system consists of a sensory organ (the olfactory epithelium) and specific olfactory brain regions and the olfactory bulb. The perception of odors causes interesting and different problems for the nervous system. Primary one is that there is no single dimension that relates stimulus to sensation. Vision and hearing are stimulated by predictable variations in frequencies of light and sound; touch by variations in frequencies of pressure on the skin. Specialized olfactory receptor cells are located in a small patch of mucous membrane lining the roof of the nose. Axons of these sensory cells pass through perforations in the overlying bone and enter two elongated *olfactory bulbs* lying on top of the bone. The portion of the sensory cell that is exposed to odors possesses hair like cilia which contain the receptor sites that are stimulated by airborne odor molecules. These molecules dissolve in the mucous lining in order to stimulate receptor proteins in the cilia to start the smell response. An odorant acts on many receptors to different degrees.

Table 5.3 Receptor family (extroreceptors)

Receptor family	Sensations that are detected
Chemoreceptor	Taste, smell
Electroreceptor	Electro-location, impedance changes, conductivity changes
Magnetoreceptor	Flight navigation
Mechanoreceptor	Vibration, pressure, strain, force, muscle, hearing
Photoreceptor	Vision
Thermoreceptor	Heat, cold

Similarly, a receptor interacts with many different odorants to varying degrees. Schematic representation of odorant path in olfactory system is shown in Fig. 5.6.

Odorant molecules have no obvious connections with each other except that they are odorous, that is, they are inducing sensations in the olfactory system. Olfactory system unique characteristic is that there is no limit to the number of odorous molecules that can be detected and described. Odorous molecules are mainly limited to molecules of 200–400 mW but within that range there are essentially an infinite number of odorous molecules. The molecular structures are highly variable and no individual or group of individuals has been exposed to all of the range, or possibly even the majority of the range. The immune system evolve to detect and respond to an open ended set of stimuli and has solved it by using a variable rearrangement of its genetic code to generate protein receptors of huge range. The olfactory system has solved this problem by generating a huge number of individual receptor genes. The olfactory system requires one thirtieth of the genome. An intriguing aspect of the olfactory system is that these odorant receptor genes are involved both in odorant detection as well as establishing the basic anatomy of the olfactory system that allows that detection [16].

Our ability to recognize, identifies, and discriminate an olfactory object across contexts and against the background of other odors can be accomplished for simple molecular compounds and not merely complex ecological odors. Though we discussed in a general way (Chap. 3: Transduction mechanism) that the firing of olfactory nerve fibers is the outcome of a non-specific increase in ionic permeability, mediated by cAMP, how odorant molecules actually trigger this change remains something of a uncertainty. Most probably there are receptor sites that recognize particular molecules or classes of molecule, but the basis of this recognition it not as straight-forward a matter as one might imagine.

The neural systems for taste and smell are distinct from one another, the feelings of flavors and aromas mostly work together, mainly during eating. Much of what normally described as flavor comes from food molecules arrived up noses. Additionally, these two senses both have connections to brain areas that control emotions, regulate food and water intake, and form certain types of memories. Connection between olfactory and gustatory systems is the constant proceeds of

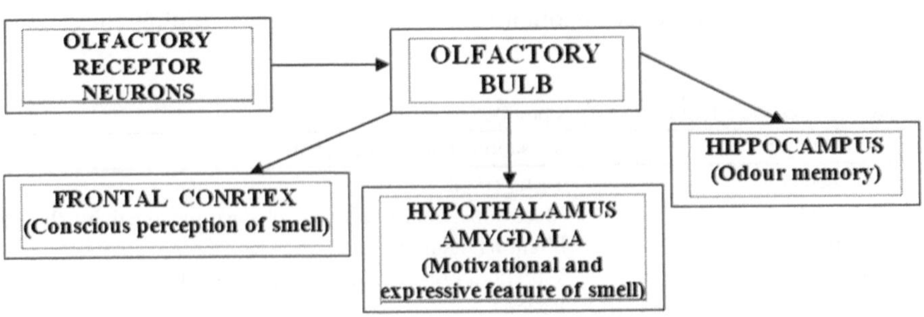

Fig. 5.6 Odorant pathways

olfactory and gustatory receptor cells. After about 10 days, taste receptor cells die and are substitute by cells that distinguish from a sort of stem cell in the taste bud. Unexpected is the story of olfactory sensory cells. These are not epithelial cells as are taste cells, but neurons, which until recently were not known to be generated in adults. The olfactory sensory neurons are not only replaced every 60 days or so, but each must also grow an axon to the correct place in the brain. Researchers are investigating how taste perception and odor recognition are maintained in the face of this turnover and new axon growth.

5.8 Signal Analysis Techniques

Human body is made up of many systems i.e., nervous system, olfactory system, cardiovascular system. Each system is made up of subsystem which sum up many physiological processes. Biomedical signal analysis is the process to deal with the signal produced by these physiological processes. These signals could be electrical, chemical or acoustic in origin and an analysis of these signals are often useful in explaining or identifying conditions of the human body. These signals are in their rawest forms do not provide much information and therefore, inspiration behind biological signal analysis is to extract the relevant information. This analysis has become more important to provide cost effective point of care diagnosis and fast treatments.

The analysis of electrical signals is a basic problem as the essential parameters of significance are frequently changed into electrical signals by means of transducers. A signal is basically the record of a process that occurs in relation to an independent variable. This independent variable can be any of a number of things, but in most cases it is time, in which case the signal is actually called a "time-series". Bio-signal contains a Physical, chemical, mechanical, thermal, electrical and magnetic quantities that provide information of health condition in physiology and psychophysiology. Transduction is a procedure by which the quantity that characterizes the property or state of an object was sensed or transduces and conditioning analogically and digitally is a procedure of obtaining wanted information or signal components from the above object quantity.

Signal and information processing/analysis technology plays an essential role in current research. Recent data acquisition equipments provide a huge amount of raw data but the valuable information is embedded within high energy noise and irrelevant surroundings activity. Therefore, advanced signal processing/analysis methods are needed to process/analyze that information before using it by the clinicians for diagnosis or therapy.

Many instruments are available for the analysis of electrical signals in the time, frequency and model domains. Any signal transducing from a biological or medical source can be called a *bio-signal*. The signal source could be at the molecular level, cell level, or a systemic or organ level.

All living things, from cells to organism, deliver signals of biological origin. Such signals can be electrical, mechanical, or chemical. All such signals can be of interest for diagnosis, for patient monitoring and biomedical research. The main task of processing biomedical signals is to filter out the signal of interest from the noisy background and to reduce the redundant data stream to only a few, but relevant parameters.

Signal analysis is a technique which includes computer algorithms to analyze and transform the signal in an effort to create natural, meaningful, and alternate representations of the useful information contained in the signal while minimize the effects of noise. In most cases signal processing is a multi-step process that involves both numerical and graphical methods.

Biomedical signal analysis is the innovative applications of biomedical signals through various creative integrations of the method and biomedical knowledge. Various applications ranges are from the construction of artificial limbs and aids for the disabled to the development of sophisticated medical monitoring systems that can operate in a noninvasive manner to give real time views of the workings of the human body. There are a number of medical systems in common use. These includes ultrasound, electrocardiography (ECG) and plythesmography are widely used for many purposes.

Biomedical signals are study of physiological activities of organisms, series from gene to protein sequences, to neural and cardiac rhythms, to tissue and organ images. Biomedical signal processing aims at extracting significant information from biomedical signals. With the help of biomedical signal processing, biologists can discover new biology and physicians can monitor distinct illnesses [17].

A fundamental method for noise cancelation analyzes the signal spectra and suppresses undesired frequency components. Another analysis framework derives from statistical signal processing.

The processing of biomedical signals usually consists of at least four stages:

- Measurement or observation, that is, signals acquisition.
- Transformation and reduction of the signals.
- Computation of signal parameters that are diagnostically significant, and
- Interpretation or classification of the signals.

Bio-signal processing stages are shown in Fig. 5.7. Basic signal processing schematic is also shown in Fig. 5.8.

5.9 Classification of Bio-signals

Biosignal may be classified in many ways according to:

- Source or physical nature: in this type classification respects the basic physical characteristics of the considered process.

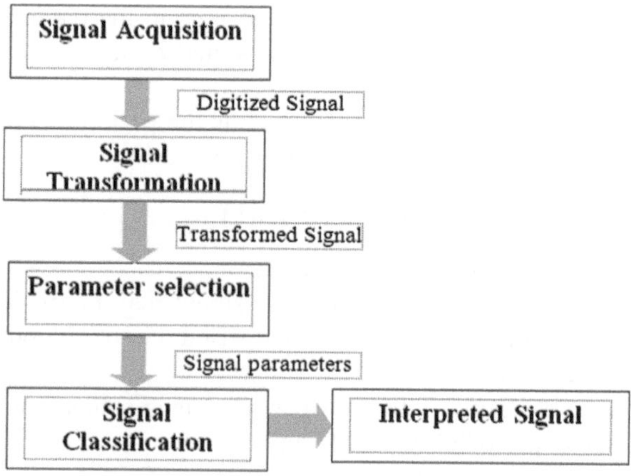

Fig. 5.7 Bio-signal processing

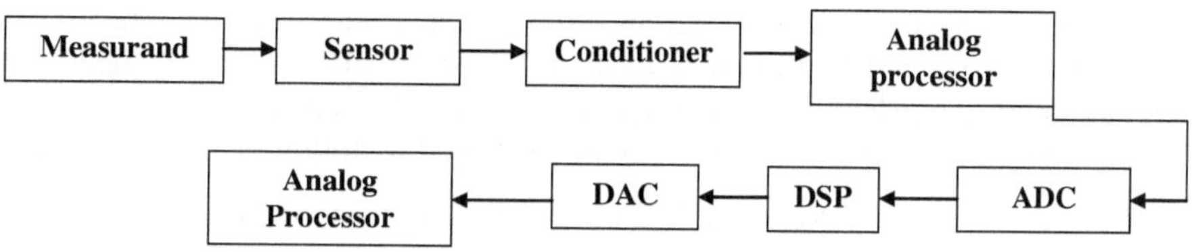

Fig. 5.8 The basic signal processing

- Biomedical application: The biomedical signal is acquired and processed with some diagnostic, monitoring or other goal in mind. Classification may be constructed according to the field of application e.g., cardiology or neurology. Such classification may be of interest when the goal is, i.e., the study of physiologic systems.
- Signal characteristic: From point of view of signal analysis, this is the most relevant classification method. When the main goal is processing, it is not relevant to the source of the signal of to which biomedical system it belongs; what matters are the signal characteristics.

In another way, biological signals can be classified into two main groups: the deterministic and the stochastic (or statistical) signals, as shown in Fig. 5.9. Such as a beating heart or respiration generates signals that are also repetitive. The deterministic group is subdivided into periodic, quasi-periodic, and transient signals. The stochastic signals are subdivided into stationary and non-stationary signals [18]. Groups of cells depolarize in a more or less random fashion such as muscle cells generating electromyography or nerve cells in cortex. Deterministic signals are signals that can be exactly described mathematically or graphically. Real-world signals are never deterministic. There is always some unknown and unpredictable noise adds some unpredictable change in the parameters. It is very often convenient to approximate or model the signal by means of a deterministic function(s).

An important family of deterministic signals is the periodic family. A periodic signal is a deterministic signal that may be expressed by

$$S(t) = s(t + \mathrm{nT})$$

Where n is an integer, and T is the period.

Under some conditions, the blood pressure signal may be modeled by a complex periodic signal, with the heart rate as its period and the blood pressure wave shape as its basic wave shape.

Most deterministic functions are non periodic. It is sometimes worthwhile to consider an 'almost periodic' type of signal. Example: The ECG signal can sometimes be considered almost periodic.

The most important class of signals is the stochastic class. Stochastic signals cannot be expressed accurately; they can be described only in terms of probabilities. Stationary stochastic processes are processes whose statistics do not change in time. The expectation and the variance (as with any other statistical mean) of a stationary process will be time-independent. Unfortunately, almost all signals are non-stationary e.g., sleep Electroencephalograph (EEG) signal.

According to origin bio-signals may be classified as follows:

1. Bioelectric signals: Nerve cells and muscle cells generate a signal which is called bioelectric signal. Membrane potential excited in certain conditions and generates action potential which is source of the bioelectric signal. A simple transducer is required for measurement of the bioelectric signal. ECG and EEGs are these types of signals which are shown in Figs. 5. 10 and 5.11.
2. Bioimpedance signals: The Bioimpedance signal is generated as the muscle/body tissue place under test sinusoidal currents with frequency range

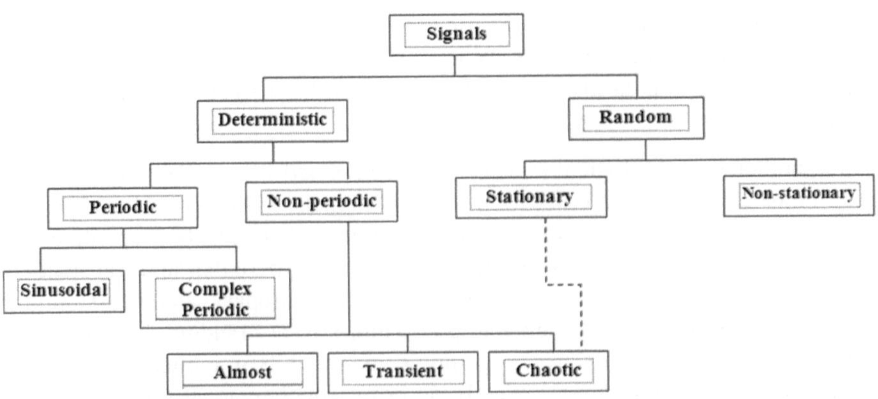

Fig. 5.9 Classification of signals

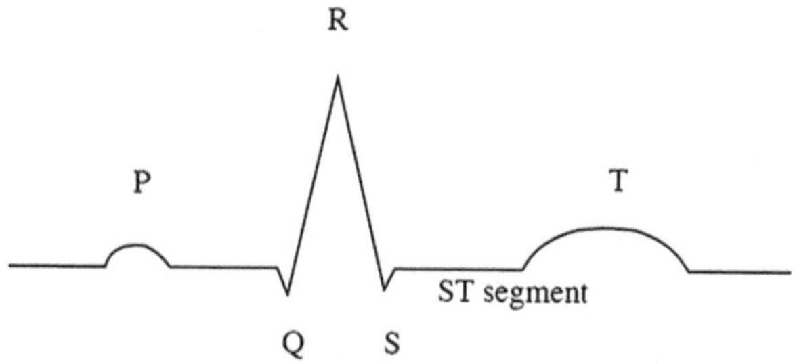

Fig. 5.10 Typical waveform of an ECG

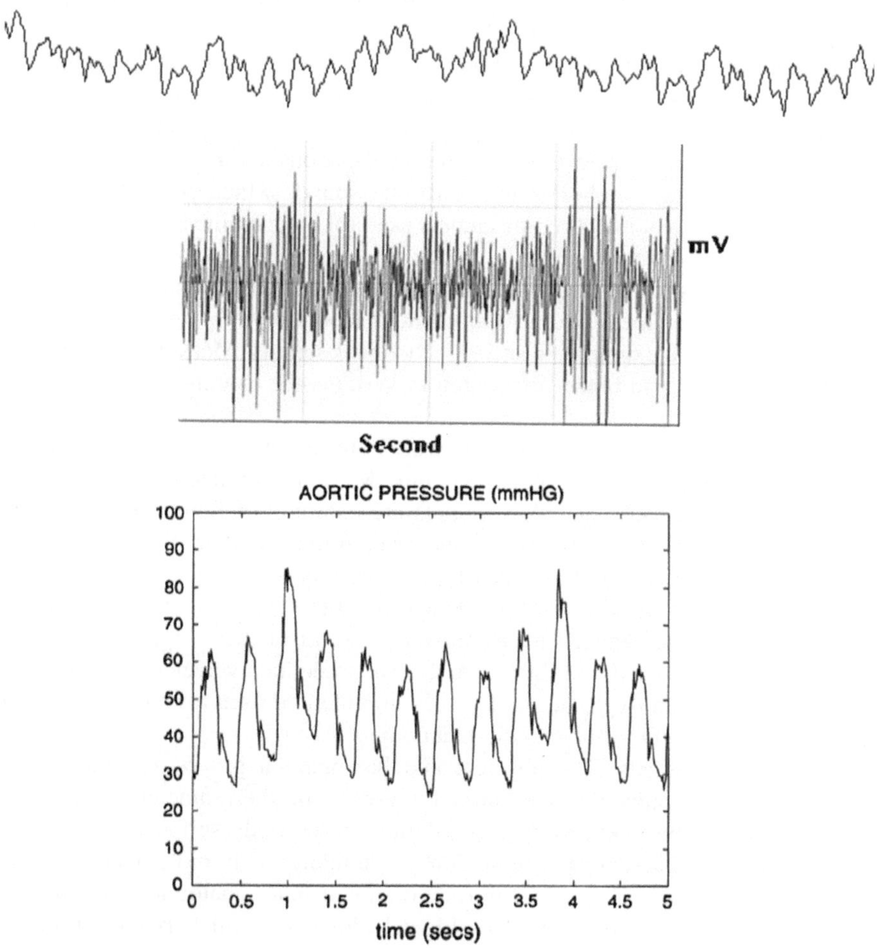

Fig. 5.11 Typical waveform of EEG, EMG and blood pressure signals

of 50 kHz–1 MHz, and low current range in between 20 μA and 2 mA. This impedance of signal contains information regarding blood volume, blood distribution, automatic nervous system activity, and more. There is a defined frequency range (50 kHz–1 MHz) chosen to minimize electrode polarization problems, and the low current range (20 μA–2 mA) are preferred to avoid tissue damage due to heating property. e.g., impedance plethysmography or measurement of galvanic skin resistance and measurement of respiratory rate based on bio-impedance.

3. Biomagnetic signals: Biomagnetic signals have very low signal to noise ratio. Various organs, such as the brain, heart and lungs, produce extremely weak magnetic fields (10^{-9}–10^{-6} T). These magnetic fields produce signals which provide information as it is useful for diagnosis and not included in bioelectrical signals. In designing this type of measuring system great caution must be taken of these signals. e.g. EEG.
4. Bioacoustics signals: Human body generates some noise or as defined in biological terms generates typical acoustic noise. i.e., the blood flow in heart through blood vessels and valve generates acoustic noise or sounds which are used to measure the pressure of the blood. Bioacoustics signal generate sounds like coughs, snores, chest sound and lung sounds which is used as primary diagnosis in medicine. The transducers used for measurement of this sound are microphones or accelerometers.
5. Biochemical signals: The cell contains concentration of ions and this ion of living tissue is measured in the clinical laboratory. So this concentration measurement is example of such biochemical signals. Measurement of blood pH, ion analysis in blood or respiratory systems are examples of biochemical signals.
6. Biomechanical signals: All signals which are the outputs of mechanical fuction occurring in the biologic system. Motion and displacement signals, pressure and tension and flow signals are example of biomehanical signals. Blood pressure, non-directly- phonocardiography, and motion of the chest wall are a good example for this type of signals.
7. Biooptical signals: The result of optical functions in biological system taking place due to the induced measurement is origin of these signals. The transmitted and backscattered light from a tissue in defined wavelengths is used for measuring blood oxygenation. Dye dilution method used for monitoring heath output is outcome of fiber optic technology.
8. Thermal biosignals: The physical and biochemical processes which occur/proceed in organism is measured by variety of thermometers. They carry information about temperature circulation on the body surface.
9. Radiological biosignals: Interaction of ionizing with biological structures formed this type of signals. Information about inner anatomical structures of human being is known by Radiological biosignals which is most useful in diagnostics and therapy.
10. Ultrasonic biosignals: Ultrasonic biosignals are produced by interaction with organ tissues. Information about acoustic impedances of biological structures

and their anatomical changes carried out by this signals. Piezo electric transducer worked as probes in this type of sensory system. e.g., Sonorgraphy machine

Characteristics of biosignals:

- *Deterministic* signal (man-made).
  - Classified by mathematical function or rules.
  - Periodic signal, transient signal.
- *Random* signal (stochastic signal).
  - Contain uncertainty in the parameters that describe it.
- Real biological signals almost always have some unpredictable noise or change in parameters.
- e.g., EMG, ECG, EEG.

5.10 Signal Analysis Techniques

Basically multivariate data analysis involves data reduction, i.e., it reduces high dimensionality in a multivariate problem where variables are somewhat correlated (e.g., sensors with overlapping sensitivities), allowing the information to be displayed in a smaller dimension (typically two or three) [19]. There are many multivariate techniques available to choose from, which is shown in Fig. 5.12. A subset, called PARC techniques, is normally used in "E-nose" data analysis.

Multidimensional scaling (MDS), principal components analysis (PCA), self organizing maps (SOM), independent component analysis (ICA), Cluster analysis (CA), linear discriminate analysis (LDA), partial least squares (PLS), feature subset selection (FSS), principal component regression (PCR), multiple linear

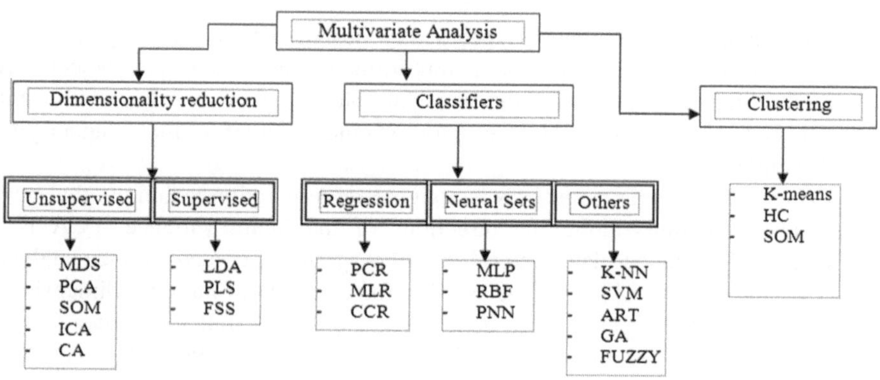

Fig. 5.12 Multivariate pattern analysis techniques methods

regression (MLR), canonical correlation regression (CCR), multilayer perception (MLP), radial basis function (RBF), probabilistic neural network (PNN), K nearest neighbors (K-NN), support vector machines (SVM), adaptive resonance theory (ART), genetic algorithm (GA), hierarchical clustering (HC).

The Pattern reorganization methods can be divided into unsupervised techniques and supervised techniques. Unsupervised learning methods are generally used in exploratory data analysis because they attempt to identify a gas mixture without prior information on the nature of the samples. These techniques, which include PCA, CA, ICA, SOM and MDS, are useful when no example of different sample groups is available, or when hidden relationships between samples or variables are suspected [19, 20].

Supervised learning techniques classify measurements by developing a mathematical model to relate training data, i.e., samples with known properties, to a set of given descriptors. Test samples are then evaluated against a knowledge base and predicted class membership is deduced. These methods enable the system to de-emphasize parameters other than volatile, for example temperature and humidity, and to train a system to look only at particular combinations of sensors [19, 20].

When high concentrations of volatile are involved, a non-linear PARC technique, such as ANN or RBF, would be more appropriate. Non-linear models usually need more parameters and therefore more input data than linear models, since some parameters are used to describe the shape of the non-linearity. The main advantage of such a method is flexibility, i.e., the ability to adjust to more complex data variations. However, caution is necessary when choosing the desired model flexibility by selecting the number of parameters. If too many parameters are taken into account, the calculated model will be over-flexible, fitting to all relevant data variations as well as to every unwanted sensor noise.

- *Dimensionality reduction*:

  A dimensionality reduction stage is required in most cases, either feature extraction or feature selection.

  Unsupervised: In unsupervised learning, the training of the network is entirely data-driven and no target results for the input data vectors are provided.

  - Multidimensional Scaling is specifically designed to graphically represent relationships between objects in multidimensional space. The objects are represented on a plot with the new variables as axes and the relationship between the objects on the plot should represent their underlying dissimilarity.
  - *Principal components analysis* is a signal representation technique that generates projections along the directions of maximum variance. PCA is a method that reduces data dimensionality by performing a covariance analysis between factors. As such, it is suitable for data sets in multiple dimensions, such as a large experiment in gene expression. PCA transforms data so that they are redefined in terms of their principal components (PCs). PCA is powerful technique for data analysis.

- The SOM was developed by professor Kohonen. The SOM has been proven useful in many applications. It is considered as one of the most popular neural network models. It belongs to the category of competitive learning networks. The SOM can be used to detect features inherent to the problem and thus has also been called the self-organizing feature map (SOFM). The SOM can thus serve as a cluster analyzing tool of high-dimensional data.
- Independent component analysis Developed extension of PCA is called *independent components analysis.* This technique is useful in arriving at an estimate of the unknown original sources of the signal. *Principal component analysis is commonly used for sorting neuronal spikes (action potentials).*
- Cluster analysis is a major technique for classifying a 'mountain' of information into manageable meaningful piles. It is a data reduction tool that creates subgroups that are more manageable than individual datum. Like factor analysis, it examines the full complement of inter-relationships between variables.

  Supervised: In supervised learning, a desired output result for each input vector is required when the network is trained.

- Linear discriminate analysis is a signal classification technique that directly maximizes class separately; generating projections where the examples of each class form compact clusters and the different clusters are far from each other. Quadratic discriminant analysis (QDA) is closely related to LDA.
- Partial least squares are the "gold standard" in chemo metrics due to its ability to handle collinear data and reduce the number of required calibration samples.
- Feature subset selection is a dimensionality reduction technique that can be used to configure small sensor arrays for specific odor-measurement applications. The goal of FSS is to find an "optimal" subset of sensors (or features) that maximizes information content or predictive accuracy. The simplest FSS approach consists of evaluating each feature individually and selecting those features with the highest scores. Unfortunately, this approach ignores feature redundancy and will rarely find an optimal subset.

• *Classifiers*:

– *Regressions*

Principal component regression is an alternative solution to the OLS (ordinary least square) co-linearity problem is to perform PCA and retain only a few of the principal components as "latent variables."
A MLR analysis is carried out to predict the values of a dependent variable, Y, when given a set of p explanatory variables (x_1, x_2,…, x_p).
Canonical correlation regression: The CCR estimator is based on a transformation of the variables in the co-integrating regression that removes the second-order bias of the OLS estimator.

– *Neural sets*

Multilayer perception: Wrapper evaluates feature subsets on the basis of their predictive accuracy on the PARC algorithm. This approach combined with *sequential forward selection* (SFS) to select features for a multilayer-perception regression problem. It has an ability to learn and generalize, smaller training set requirements, fast operation, ease of implementation and therefore most commonly used neural network. Experimental results showed that their feature selection procedure could find small (5–10) feature subsets with similar or better predictive accuracy than the complete set of 49–85 features.

Radial basis function: RBF networks are used for EEG signal classification. Because the networks train rapidly, usually orders of magnitude faster than MLP, while exhibiting none of its training pathologies such as paralysis or local minima problems.
Probabilistic neural network: PNN is closely related to Parzen window pdf estimator. A PNN consists of several sub-networks, each of which is a Parzen window pdf estimator for each of the classes.
– *Others*

K nearest neighbors is a classical classification technique widely used in pattern reorganization problems, to determine item class. This method investigates its neighborhood class. K-NN is a type of instance-based learning, or lazy learning where the function is only approximated locally and all computation is deferred until classification.
Support vector machines have very good solid foundation in statistical learning theory, and guarantees to find the optimal decision function for a set of training data, given a set of parameters determining the operation of SVM.
Adaptive resonance theory networks were designed to address the stability–plasticity dilemma, are capable of real-time learning and classification have been applied with some success to EN data.
Genetic algorithm, conversely, is inspired by the process of natural selection and performs a global random search on a population of solutions.
FUZZY: The theory of fuzzy logic attempts to enable machines to deal with imprecise language used by humans in order to describe data that may not be exact or crisp. Fuzzy-based computational algorithms are attractive to researchers in machine olfaction due to several sources of fuzziness can be identified in the recognition of olfactory signals such as noisy data, imprecise measurement and odorant sample sets which overlap in the feature space.

Discriminate function analysis (*DFA*), in a similar manner to PCA, also transforms data using linear discriminate functions (LDFs).

- *Clustering:*
  Clustering is an unsupervised learning process, search to discover spatial relationships or similarities among data samples, which may be hard to discern in high-dimensional feature space. The process of clustering involves three basic steps: (1) defining a dissimilarity measure between examples, typically the

Euclidean distance, (2) defining a clustering criterion to be optimized, typically based on within- and between-cluster structure (e.g., elongated, compact or topologically-ordered clusters), and (3) defining a search algorithm to find a "good" assignment of examples to clusters, since exhaustive enumeration of all possible clustering is clearly impractical [21].

- Hierarchical Clustering: These algorithms are capable of generating a multi-level clustering using a tree structure known as a dendrogram. These dendrograms can be constructed in a bottom-up or top-down method.
- Self-Organizing Maps: SOMs are connectionist techniques capable of generating topology-preserving clustering [22]. An SOM is a network of clusters (or neurons) arranged in a lattice structure, typically two-dimensional. The behavior of SOMs result from the synergy of three processes: competition, cooperation, and adaptation [23]. First, all neurons in the lattice enter a *competition* for each incoming example. The closest neuron in feature space is selected as a winner and becomes activated. SOMs have very interesting properties for data visualization but mapping onto the SOM manifold can be tricky if the structure of the data is inherently high dimensional.
- C means: C-means is a clustering algorithm that generates a single-level partition of the dataset. Starting from an initial clustering (e.g., a random assignment of examples to clusters), C-means iteratively re-computes the sample mean of each cluster and reassigns each example to the cluster with the closest mean. The basic C-means algorithm requires a pre-specified number of clusters, heuristic procedures [24] can be employed to automatically determine an appropriate number of clusters.

References

1. D.T. Win, The electronic nose—a big part of our future. AU. J. T. **9**(1), 1–8 (2005)
2. R. Haddad, A. Medhanie, Y. Roth, D. Harel, N. Sobel, Predicting odour pleasantness with an electronic nose. PLoS Comput. Biol. **6**(4), e1000740 (2010). doi:10.1371/journal.pcbi.1000740
3. M. Bernardine Dias, Investigating the viability Of MEMS vapor sensors for detecting land mines, CMU-RI-TR-00-24, Carnegie Mellon University, Pittsburgh, Pennsylvania 15213, October 2000
4. J.W. Gardner, P.N. Bartlett, *Electronic Noses. Principles and Applications* (Oxford University Press, Oxford, 1999), pp. 221–245
5. AD Wilson et al., Applications and advances in electronic-nose technologies. Sensors **9**, 5099–5148 (2009)
6. M. Ghasemi-Varnamkhasti et al., J. Food Eng. **100**, 377–387 (2010)
7. K. Toko, Electronic tongue. Biosens. Bioelectron. **13**, 701–709 (1998)
8. C. Di Natale, E. Mazzone, A. Mantini, A. Bearzotti, A. D'Amico, A.V. Legin, A.M. Rudnitskaya, Y.G. Vlasov, Electronic tongue distinguishes different mineral waters. Alta Frequenza Rivista Di Elettronica **11**(2), 88–90 (1999)

9. F. Winquist, P. Wide, I. Lundstom, An electronic tongue based on voltammetry. Anal. Chim. Acta **357**, 21–31 (1997)
10. A. Riul Jr, R.R. Malmegrim, F.J. Fonseca, L.H.C. Mattoso, An artificial taste sensor based on conducting polymers. Biosens. Bioelectron. **18**(11), 1365–1369 (2003)
11. K. Toko, Taste sensor with global selectivity. Mater. Sci. Eng. **4**(2), 69–82 (1996)
12. F. Winquist, C. Krantz-Rülcker, I. Lundström, Electronic tongues and combination of artificial senses. Sensors Update **11**(1) 279–306 (Wiley, 2002)
13. A. Menini, L. Lagostena, A. Boccaccio, Olfaction: from odourant molecule to the olfactory cortex. New Phys. Sci. **19**, 101–104 (2004)
14. H.T. Nagle, S.S. Schiffman, R. Gutierrez-Osuna, The how and why of electronic noses. IEEE Spectr. **35**(9), 22–31 (1998)
15. D.E. Hornung, M.P. Enns, Possible mechanisms for the processes of referred taste and retronasal olfaction. Chem. Senses **11**, 616 (1986)
16. M.-S. Alan, J.-P. Royet, Structure and function of the olfactory system in *Olfaction and the Brain*, ed. by W. J. Brewer et al., (Cambridge University Press, 2006)
17. H.-H. Chang, J. M. F. Moura, Biomedical signal processing (Chapter 22) in *Biomedical Engineering and Design Handbook*, ed. by Myer Kutz (McGraw Hill, 2009)
18. J. Van Bemmel, M. Musen, *Handbook of Medical Informatics*, 2nd edn. (Springer, Houten Diagem, 1997)
19. J.W. Gardner, P.N. Bartlett, A brief history of electronic noses. Sens. Actuators B **18**(1–3), 211–220 (1993)
20. J.W. Gardner, P.N. Bartlett, Pattern recognition in odour sensing, in *Sensors and Sensory Systems for an Electronic Nose*, ed. by J.W. Gardner, P.N. Bartlett (Kluwer Academic Publishers, Dordrecht, 1992), pp. 161–179
21. R. Gutierrez-Osuna, Pattern analysis for machine olfaction: a review. IEEE Sens. J. **2**(3), 189–202 (2002)
22. T. Kohonen, Self-organized formation of topologically correct feature maps. Biol. Cybern. **43**, 59–69 (1982)
23. S. Haykin, *Neural Networks, a Comprehensive Foundation*, 2nd edn. (Prentice-Hall, Englewood Cliffs, 1999)
24. C.W. Therrien, *Decision, Estimation and Classification: An Introduction to Pattern Recognition and Related Topics* (Wiley, New York, 1989)

Chapter 6
Sensor Used in E-nose

6.1 Introduction

A sensor is a device that is used to measure physical quantity, and the opposite device is the actuator, which converts signal into action and in transducer one form of energy is electrical and therefore actuator and sensor are the other forms of the transducer. Sensors provide us with a means of generating signals that can be used as inputs to electronic circuits. Biosensors are analytical tools for the analysis of biomaterial samples to gain an understanding of their biocomposition, structure, and function by converting a biological response into an electrical signal. In contrast with conventional bioarrays, biosensors allow the detection of molecular interactions as they take place, without requiring auxiliary procedures, making them highly attractive for biotechnological applications.

6.2 Measurement of Smell

Multiple useful roles are played by sensations of smell. It influences emotional states, i.e., enthusiasm or attention. The sensation of smell allows the control of food purity by word of warning beside bad food after remembering its association with unpleasant odors; it also provides information about the family, or outsiders in animals without skill of language; and last but not least it also activates salivary and gastric emissions in response to pleasant odors. Unpleasant odors can evoke symptoms.

Symptoms evocation, concentration disturbance, and productivity reduction are the examples of unpleasant odors in the environmental air. Number of molecules available in the concentration of the olfactory receptors is defined as odor sensation dependence.

In the air, only volatile substances which may be associated with particular matter have an odor. Sense of smell is capable to distinguish thousands of odors but it is very difficult to categorize them based on their chemical structure

H. K. Patel, *The Electronic Nose: Artificial Olfaction Technology*, Biological and Medical Physics, Biomedical Engineering,
DOI: 10.1007/978-81-322-1548-6_6, © Springer India 2014

(molecular shape). A trained human can distinguish several classes of compounds, i.e., ethereal (wine), aromatic (camphor, menthol), balsamic (violet, vanilla), alliaceous (garlic, iodine), ambrosial (amber, musk), empyreumatic (coffee), which classify an acrid odor constricted by an organic matter when subjected to the action of a strong heat, sweat, repulsive (pyridine, opium), or nauseous (feces, putrefaction) [1].

Molecules of various chemicals in the air are outcome of smell. Substance which contains chemicals and volatile in nature has smell. Human nasal cavity contains millions of neural receptors of 350 different types. Response to different chemicals depends on type and quantity of different sets of neural receptors and they are activated and then after brain decodes each pattern and assign meaning to it. Sense of smell varies among persons and also can decrease with age. Odors blend when mixed, making individual difficult to recognize.

Hundreds of molecules from the cake, sandwich, and juice are mixed together in the air, but the person just perceives "Cake," "Sandwich," and "Juice." This perception of three odors from hundreds of intermixed molecules is a feat of perceptual organization (Fig. 6.1).

There are two traditional methods for odor or SMELL analysis used in various industries (Sensory evaluation and Instrumental evaluation: discussed in Chap. 4). One method is to use advanced analytical instruments in the laboratories, i.e., instrumental evaluation. Instrumental evaluation techniques can give especially detailed information about the specific molecules contents of the smell. These classical analytical techniques involve GC–MS, LC–MS, HPLC, HPTLC, etc., that can separate, identify, and quantify individual chemicals. Odors are usually composed of a complex mixture of different volatiles; such techniques are too painful for practical in everyday applications and costly to set up. Also many volatile chemicals are in very small quantities and beyond their detection limits. Additionally, the relationship between the physical and chemical properties of the odorant molecules and their sensory impact is still undergoing research.

The other traditional method is the use of human test panels. Human sensory evaluation is a powerful method. The human nose has been playing an important role in assessing the quality of many products for a long period of time.

Fig. 6.1 Human perception of odor

Traditionally, expert human panels are employed for this purpose. However, reliance on human nose for VOC detection is difficult. If something is hazardous, odor itself is an indication of same. Many times hazardous chemicals have no odor or some would have pleasant odor while some substances have offensive odor. A specific substance smell concentration level can only be smelled by human nose. Human nose can easily come into a state of "olfactory exhaustion."

Human nose can only smell a substance beyond a certain concentration level. Below this odor threshold, VOC can no longer be perceived. Even for a concentration beyond the odor threshold, human nose can easily come into a state of "olfactory fatigue": human nose comes to a point that it is no longer able to smell the odor while the odor still maintains a high concentration level. Due to these limitations, a lot of studies have been performed trying to use chemical sensors to replace human nose for odor detection.

A different way to independently analyze an odor is to design an instrument to mimic the human sense of smell. This alternative technology will complement or in some cases will be used as a replacement of the currently used techniques. This process development is useful for the recognition system to produce unique classification system of each chemical or smell, so that an automated identification of that chemical or smell can be applied. The method to record or mimic electronically the human olfaction sense is characterized by inadequate and very preliminary approaches.

6.3 Analysis and Classification of Sensor

All chemical sensors comprise appropriate, chemically sensitive materials that are interfaced to a transducer. Interaction of the analyzed molecules with the chemically sensitive material generates some physical changes that are sensed by the transducer and converted to an output signal. The range of gas-sensing materials is potentially very broad and can be divided into a number of ways, either by material type or by the nature of the interaction with the analyzer [2]. Table 6.1 summarizes the biological and engineered sensors for respective functions.

Table 6.1 Examples of engineered and biological sensors [30]

Engineered sensors		Biological sensors	
Sensor	Sensed quantity	Sensor	Sensed quantity
Photodiode	Light intensity	Retina	Light intensity
Psychrometer	Humidity	Chochlea	Sound
Barometer	Pressure	Ear canal	Level, rotation
Thermometer	Temperature	Taste bud	Chemical composition
Phenolphthalein	pH	Skin	Temperature
Timer	Time	Skin and hair	Air flow
Odometer	Distance(inferred)	Olfactory cells	Gas composition

These interactions are dependent on the shapes and the charge distributions within the analyzed molecules and the sensor materials, and are similar to the interactions operative in the biological system between the odorants and the receptor proteins. The types of odor sensors that can be used in an E-nose require to respond to odorous molecules in the gas phase.

A device that transforms information, ranging from the concentration of a specific sample component to total composition analysis, into an analytically useful signal is defined as a chemical sensor. Two main basic components connected in series and implemented in chemical sensors are usually: a chemical (molecular) identification system (receptor) and physiochemical transducer.

A chemical sensor in which the recognition system utilizes a biochemical mechanism is defined as biosensors [3, 4]. A biosensor is a chemical-sensing device in which a biologically derived recognition entity is attached to a transducer to allow the quantitative development of some complex biochemical parameters [3] or biosensors detect molecules with high selectivity on the basis of molecular recognition. Biosensors can be grouped according to their biological or transduction elements [5, 6]. Biological elements include enzymes, antibodies, microorganisms, biological tissue, and organelles [7]. Biosensors are sometimes defined as analytical devices incorporating either a biological material, such as a tissue, microorganism, organelle, cell receptor, enzyme, antibody or nucleic acid; a biologically derived material, for example recombinant antibody, engineered protein; a biomimic such as a synthetic catalyst, combinatorial ligand, or imprinted polymer. Biosensors usually yield a digital electronic signal, which can be related to the concentration of a specific analyte or group of analytes.

Biosensors are electrical, optical, chemical, or mechanical devices with the capability to detect biological species selectively. They are often modified with biological entities to enhance their selectivity. Examples of biological recognition molecules include enzymes, antibodies, and oligonucleotides. The ideal biosensor not only has to respond to low concentrations of analytes, but also must have the ability to discriminate among species according to the recognition molecules that are immobilized on its surface. Biosensors have wide applications which include biomarker detection for medical diagnostics, pathogen, and toxin detection in food and water.

Biosensor molecules detect signal using a transducer, e.g., optical, electrochemical, thermal, piezoelectric, and gravimetric transducers are used to get signals in these devices. Fundamental system of detection is the signal transduction associated with the careful identification of biological or chemical species [4, 8]. There are two different types of biosensors such as biocatalytic- and bioaffinity-based biosensors. The biocatalytic biosensor uses mainly enzymes as the biological compound, catalyzing a signaling biochemical reaction [9].

A biosensor which is designed to monitor the binding event itself uses specific binding proteins, lectins, receptors, nucleic acids, membranes, whole cells, antibodies, or antibody-related substances for biomolecular recognition is basically defined as bioaffinity sensor [5–7].

A chemical biosensor is a sensor that produces an electrical signal proportional to the concentration of biochemical analytes. These biosensors use chemical as well as physical principles in their operation [10, 11]. Interdisciplinary studies in chemistry, electronics, and biology have led to the development of novel sensor interfaces. Based on unique physical, chemical, and electrocatalytic properties, nanoparticles play variety of roles in different biosensing systems. The attachment of nanoparticles onto electrodes drastically enhances the conductivity and electron transfer from the redox analytes to make them electroanalytical sensor [12].

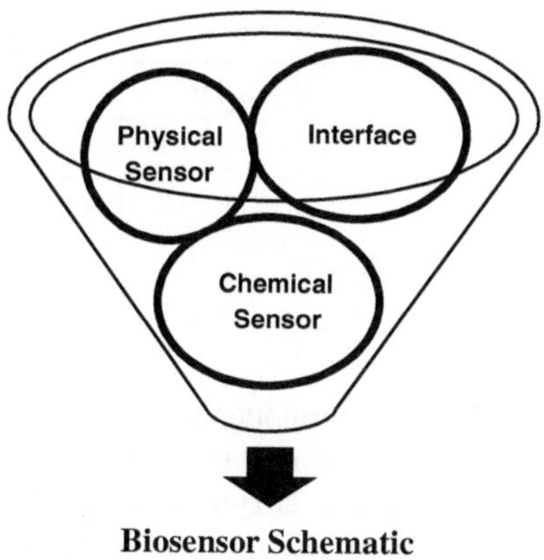

Figure 6.2 describes a typical biosensor configuration that allows measurement of a target analyte without using reagents. The device incorporates a biological-sensing element with a traditional transducer.

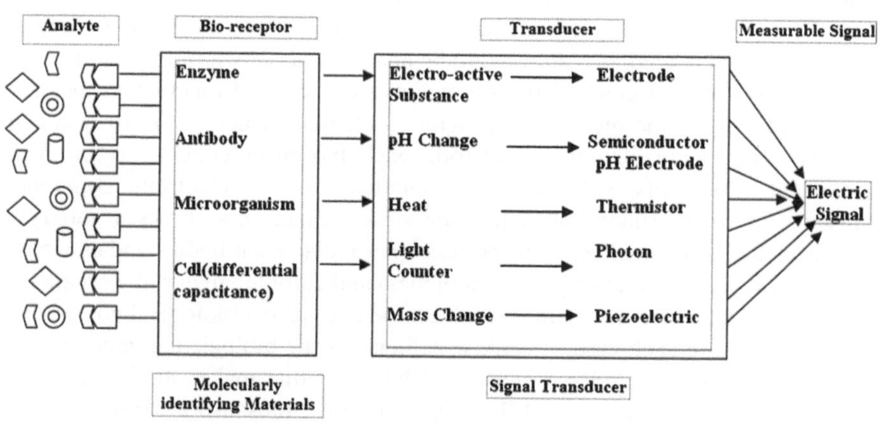

Fig. 6.2 A biosensor configuration

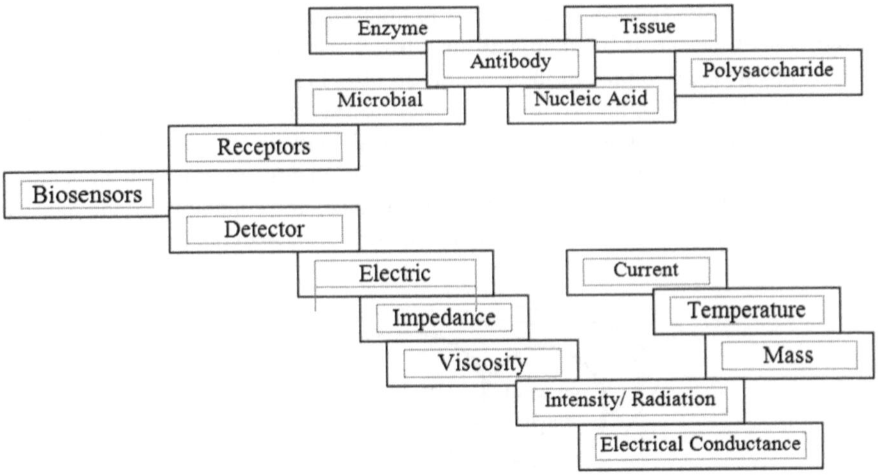

Fig. 6.3 Biosensor elements

Basically, a biosensor has two key components such as a receptor and a detector (as shown in Fig. 6.3). The receptor is dependable for the selectivity of the sensor. Examples of common receptors include enzymes, antibodies, nucleic acids, and lipids. The detector plays the role of the transducer, translates the physical or chemical change occurrence at the receptor by recognizing the analyte and communicated it through an electrical signal. As a rule, then, the detector component of a biosensor is not selective. Examples of transducer elements are ranged from pH and oxygen electrodes to piezoelectric crystals.

Biosensors required suitable transducers to obtain measurable signal for the detection of desired analyte. There are various combinations of the biological material and transducers that are available, which depend on each sample of interest and the type of physical magnitude to be measured. In biosensor system, there are number of possible analytes, required to be analyzed as shown in Fig. 6.4 [13].

Biosensors are self-controlled parts which are packaged together in the same unit, usually small, the biological recognition element which is in direct contact with the transducing element. Therefore, each biosensor contains a biological component, which acts as the sensor and another one is an electronic component which works to transduce and detect the signal. Figure 6.4 shows a variety of substances including nucleic acids, proteins (particularly antibodies and enzymes), lectins (plant proteins that bind sugar moieties) and complex materials (organelles, tissue slices, and microorganism), which can be used as the biological components or bioreceptors. In each case, it is the specificity of the biological components for an analyte (or group of related analytes: LIGAND) that makes the biomolecules smart as sensing component. Like other sensors, biosensors "transduce" or translate one kind of energy into another and must be placed into a chain of

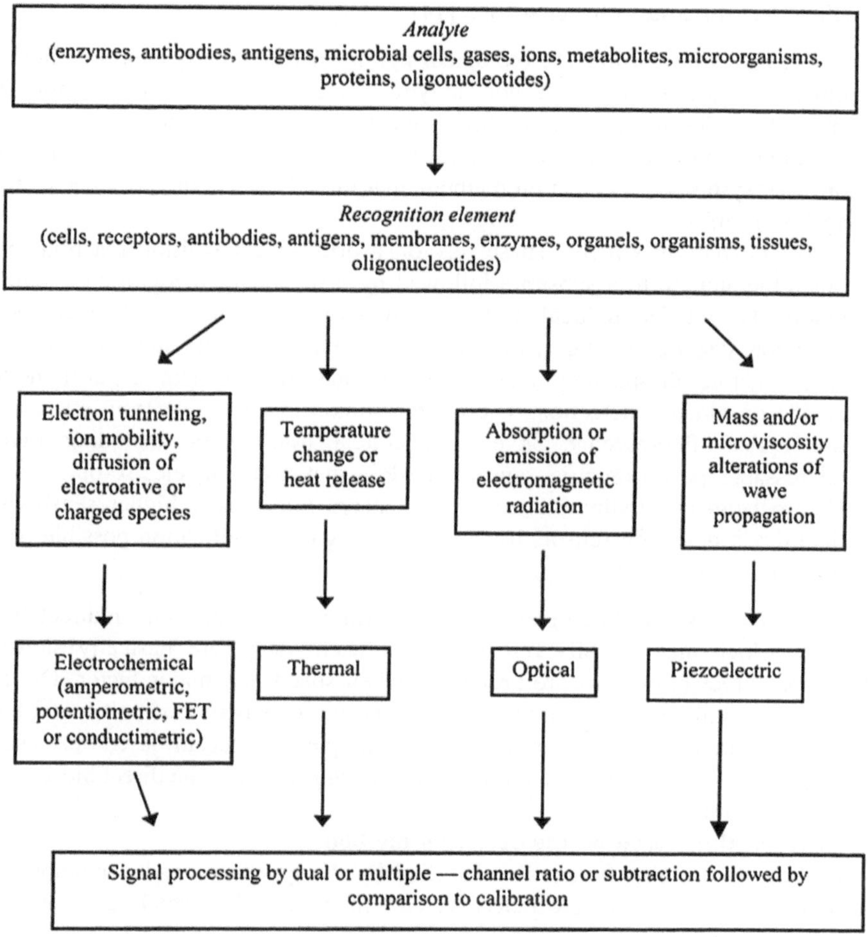

Fig. 6.4 Bio component and transducers in a biosensor system [13]

components that detect, translate, and modify an original "indication from being" into a useful signal.

Both components of the biosensor, namely, biorecognition element (referred as a receptor) and transduction platform (referred as a transducer) play an important role in the construction of a sensitive and specific device for the analyte of interest (referred as a target).

There are numerous ways biosensors are classified:

1. Direct sensor and indirect sensor.
2. Classification based on the detection method.
3. Classification based on the method of transduction.
4. Classification based on the mode of delivery.

Direct sensor and indirect sensor [14]

1. *Direct-detection sensors* The biological significance is directly measured in real time, and they are typically noncatalytic elements such as cell receptors or antibodies. The most common direct-detection biosensor systems use surface Plasmon resonance (SPR), resonant mirror, or quartz resonator transducers. All of these systems use detection methods that directly sense the presence of the analyte itself.
2. *Indirect-detection sensors* These rely on a primary recognition reaction that binds the analyte to a substrate followed by a secondary recognition reaction that binds a labeled molecule to the joined analyte to allow detection. Indirect-detection systems use fluorescent labels, catalytic elements such as enzymes (e.g., alkaline phosphates), microorganisms, and plant or mammalian tissue to enable or enhance detection. The most common indirect-detection biosensors use optical fluorescence or electrochemical transducers, including light-addressable potentiometric sensors (LAPS) that combine electro-optics and electrochemistry. Indirect-detection biosensors often show higher sensitivity than direct-detection sensors because of the signal amplification possible with the classification step.

Direct biosensor technologies enable real-time detection, while indirect biosensors, which are typically rely on a sandwich assay, are basically slower; therefore, direct biosensors are preferable when speed is a major factor. On the other hand, indirect biosensors are likely to be more sensitive, simpler, and less expensive than many direct biosensors. For example, fluorescent-based biosensors are low priced to configure in a multichannel design, whereas the direct biosensors typically are more costly.

Classification based on the detection method

Biosensors are also frequently classified by the type of transducer employed (electrochemical, optical, piezoelectric, Gravimetric, and thermal) as shown in Fig. 6.5.

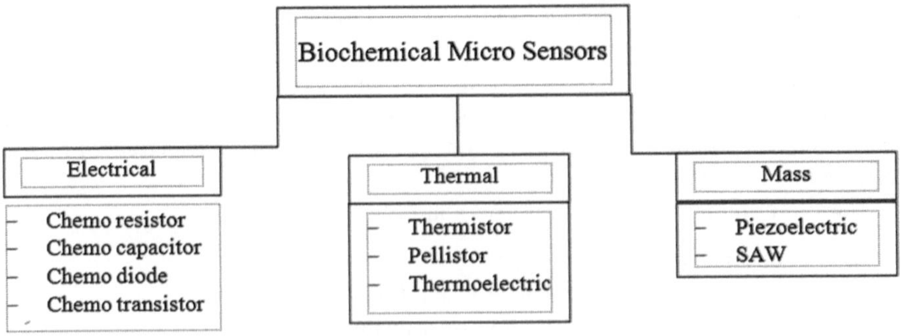

Fig. 6.5 Classification of biosensor (detection method)

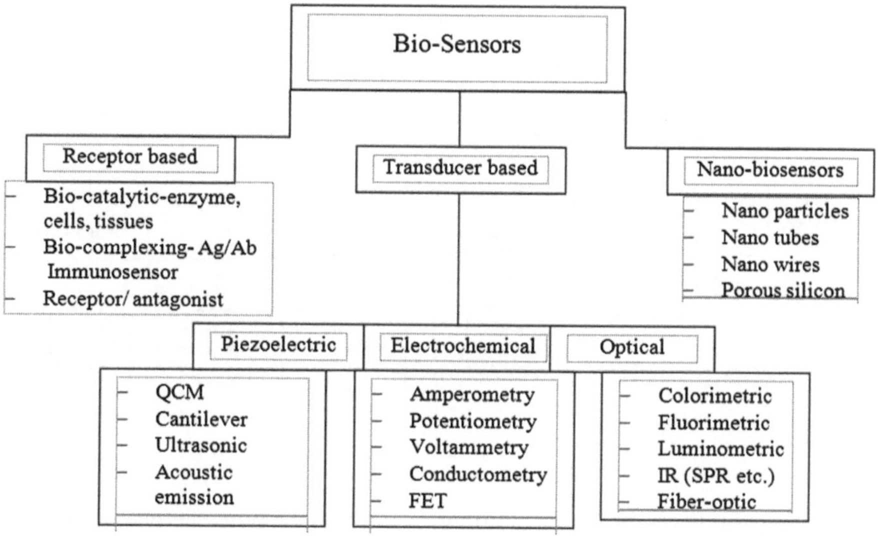

Fig. 6.6 Classification of biosensor (transduction base)

Classification based on the method of transduction

This mainly works on the principle "detection of the steady state concentration of the species formed, lost or inhibited by the transducer." Biocatalysts used by the type of biorecognition element utilized (antibody, enzymes, nucleic acids, and whole cells) as shown in Fig. 6.6.

Classification based on the mode of delivery

1. *Continuous/Flow mode* They involve the uninterrupted measurements of the analyte. Reactions take place during the flow of samples or biorecognition element. Pumps are normally used to flow samples.
2. *Intermittent/Batch mode* They involve the introduction of the analyte into the flow stream.

Ideal Biosensor Characteristics

1. Sensitivity: high $\Delta S/\Delta c_{\text{analyte}}$ (S = signal).
2. Easy calibration (with standards).
3. Linear Response: $\Delta S/\Delta c_{\text{analyte}}$ constant over large concentration range.
4. Background Signal: low noise, with ability for correction (e.g., 2nd fiber sensor head lacking biological species to measure background O_2 changes).
5. No hysteresis—signal independent of prior history of measurements.
6. Selectivity—response only to changes in target analyte concentration.
7. Long-term stability—not subject to fouling, poisoning, or oxide formation that interferes with signal; prolonged stability of biological molecule.
8. Dynamic response—rapid response to variation in analyte concentration.

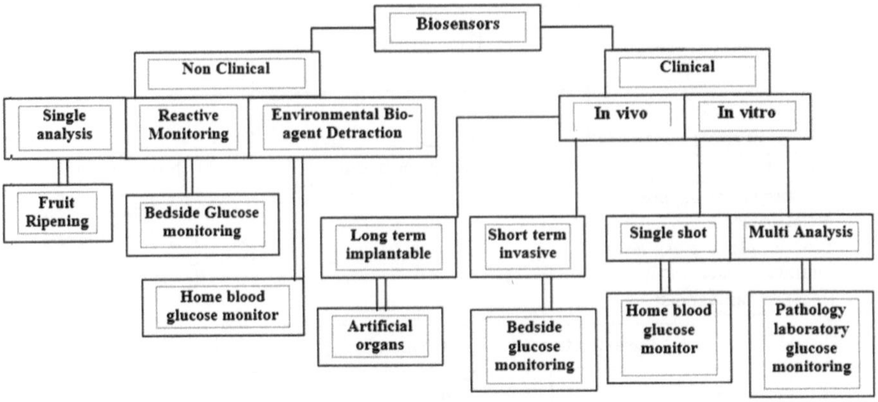

Fig. 6.7 Application of biosensor

9. Biocompatibility—minimize clotting, platelet interactions, activation of complement when in direct contact with bloodstream.

Application of biosensor

There are many applications available with various types of biosensors. The critical requirements of a biosensor include for research and commercial applications, which are required for the identification of selected molecules, i.e., glucose monitoring in diabetes patients, detection of pesticides in water contaminants, detection of pathogens, determining levels of toxic substances before and after bioremediation, etc. Some major applications are classified in Fig. 6.7.

6.4 Types of Sensor Used in E-nose

Selection of biosensor technology relies on many factors, which can be divided into three groups:

1. The nature of the analyte to be identified;
2. The nature of the sample evaluation;
3. The principle and the condition of the analysis.

All types of sensors reveal communications with the gas to be measured when volatile compounds flow over the sensor and due to that a series of physical and chemical interactions are established.

The ideal sensors to be added in an electronic nose (E-nose) should accomplish the following criteria:

- High sensitivity to chemical compounds, i.e., similar to the human nose near to 10^{-12} g/mL.
- Low sensitivity designed for humidity and temperature.

- Medium selectivity, i.e., respond to special compounds present in the sample.
- High stability.
- High reproducibility and reliability.
- Short response and recovery time.
- Dynamic and longlasting.
- Easy to calibrate.
- Easily processable data output.
- Small size.

Number of types of sensors are available and used as E-nose sensors for the analysis of volatile gases, which includes metal oxide semiconductors (MOS), conducting polymers (CP), chemocapacitors; MOS field effect transistors (MOS-FET), quartz crystal microbalance (QCM), surface acoustic wave (SAW), and SPR. The sensors which are used to detect the molecules of chemicals are based on the measurement principles such as electrical, thermal, optical, and mass changes.

6.5 Biochemical Sensors

A chemical sensor is defined as a device that transforms chemical information, ranging from the concentration of a specific sample component to total composition analysis, into an analytically useful signal. Odor molecules based on the reaction between the odor molecules and the objective sensing materials on the sensor surface detected by chemical biosensors. These reactions activate a positive change in physical properties, i.e., mass, volume, or other. This change is converted into an electronic signal by a transducer.

The combination of bio and electrochemistry, solid-state and surface physics, bioengineering, data processing implies the possibility of a new generation of highly specific, sensitive, selective, and reliable biochemical sensors.

Different types of transducers are used in chemical sensors: optical, electrochemical, heat-sensitive, and mass-sensitive. This transducers principles also work as biochemical sensor as described in Table 6.2.

Table 6.2 Principle, measurand and typical examples of biochemical sensor

Principle	Measurand	Typical sensor
Conduct metric	Resistance/conductance	Tin oxide gas sensor
Potentiometric	Voltage/e.m.f.	Ion selective FET for pH
Capacitive	Capacitance/charge	Polymeric humidity sensor
Amperometric	Current	Electrochemical cell
Calorimetric	Heat/temperature	Pellistor gas sensor
Gravimetric	Mass	Piezoelectric or SAW sensors
Optical	Path length/absorption	Infra-red detector for Methane gas
Resonant	Frequency	Surface Plasmon
Fluorescent	Intensity	Fiber optic

6.6 Vapor/Odor Sensor Classification (E-nose Sensor Classification)

1. Optical

Its basic principles not based on electrical changes in resistance, potential, current, or frequency, but the modulation of light properties is measured. Optical instruments are more complex but offer a variety of different measuring possibilities. Different operation modes were developed and are deployed using changes in adsorbance, fluorescence, optical layer thickness, and polarization.

(a) *Flourscence* This type of chemo sensor consists of optical fibers deposited with fluorescent indicator Nile Red dye in polymer matrices of varying polarity, hydrophobicity, pore size, elasticity, and swelling tendency to create unique sensing regions that interact differently with vapor molecules [15].
(b) *Absorption* Moisture-related *energy absorption* of near-infrared light can be used for measuring the moisture content of solids, liquids, and gases. Optical light absorption sensor sends light into the material of interest (medium) and makes use of the accurately measured absorption of light to determine the composition of medium.
(c) *Reflective* A reflective optical sensor consists of a housing containing both a light source and a detector. The source radiates light to outwards and it is reflected back if an object passes in front of the sensor. The reflected light is sensed by the detector.

2. Thermal Sensor

Pellistors or calorimetric sensors (subtype of thermal sensor) have not been generally in use for E-nose systems. The basic design of these sensors is a catalytic surface; platinum and palladium are most common with a heater to maintain the sensors operating temperature as well as a temperature probe. The temperature changes detected during sensing are related to the type of gas and gas concentration of the combustible analyte. Pellistors, thermistors, and thermopiles are types of thermal sensors, which measure heat changes due to chemical reactions combustion, adsorption, and enzymatic reactions/combustion. Any chemical reaction or even absorption/desorption process releases or adsorbs from its surroundings a certain quantity of heat. Figure 6.8 shows the type of sensor (pellistor) and odor sensor (thermal type) used as one of bridge elements for analysis of molecules elements.

3. Electrochemical

In the area of sensor technology, the word “chemical sensor” is defined as a group of sensors that are basically different to other sensors, i.e., thermal, magnetic, optical, and mechanical sensors. Devices that convert a chemical state into an electronic signal are chemical or biochemical sensors. The detection method for electrochemical biosensors involved measurement of current, voltage, conductance, capacitance, and impedance. Chemiresistive, potentiometric, and amperometric are this type of sensors.

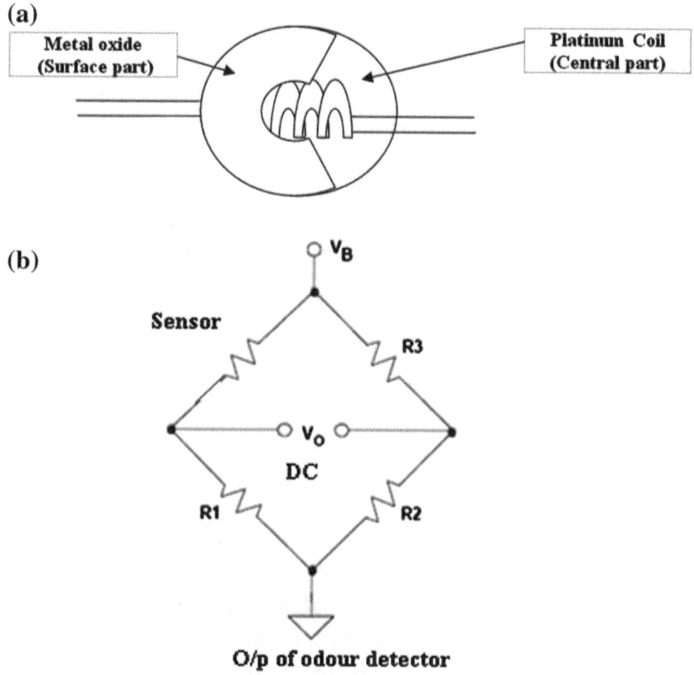

Fig. 6.8 **a** Pellistor sensor, **b** odor sensor used in wheatstone bridge

(a) Chemiresistive

The working principle of this material is that a change in property of the material resulting from interaction with a gas/odor leads to a change in resistance in the sensor. For each material type, the mechanisms that lead to resistance changes are different. Conducting polymer (CP), intrinsically CP and metal oxides (MOS) are three of the most commonly utilized classes of sensing materials. Figure 6.9 shows the MOS type chemiresistive sensor used in E-nose system.

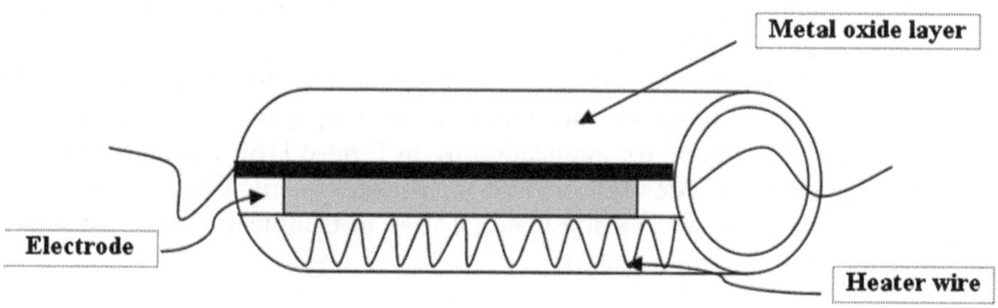

Fig. 6.9 Chemiresistive sensor

(b) Potentiometric: MOSFET

In potentiometric sensors, the zero-current potential (relative to a reference) developed at a selective membrane or electrode surface in contact with a sample solution is related to analyte concentration. The main use of potentiometric transducers in biosensors is as a pH electrode. Biosensors based on ion-selective electrodes (ISE) and ion-sensitive field effect transistors (ISFET) are defined as potentiometric biosensors. Potentiometric biosensors primary output signal is might be due to ions gathered at the ion-selective membrane interface.

(c) Amperometric

The amperometric gas sensor (AGS) was one of the initial sensors to be used in an E-nose design and has been included in a various sensor array-based instrument. Amperometry is an old electroanalytical technique that encompasses coulometry, voltametry and constant potential techniques and is widely used to identify and quantify electroactive species in liquid and gas phases. Application of amperometry to gas phase analytes involves a unique gas–liquid/solid interfacial transport process. The common characteristic of all AGSs is that measurements are made by recording the current in the electrochemical cell between the working and counter electrodes as a function of the analyte concentration. An amperometric sensor consists of a working, counter, and reference electrodes that are dipped in an electrolyte. The analyte molecules diffuse into the electrochemical cell and to the working electrode surface through a porous membrane. Then, the analyte reacted electrochemically, i.e., oxidized or reduced, and this process, governed by Faraday's Law, either produces or consumes electrons at the working electrode. The amperometric class of electrochemical sensor complements the other two classes of electrochemical sensors, i.e., potentiometric sensors that measure the null potential at zero current, and conductometric sensors that measure changes in impedance.

4. Gravimetric

As one of the most important sensing principles, gravimetric sensor systems are often used for E-noses. The main subgroups are bulk acoustic wave (BAW), surface acoustic wave or sound acoustic wave (SAW) or flexural plate wave (FPW), or shear horizontal acoustic plate mode sensors (SH-APM). BAW and SH-APM are commonly referred as thickness shear mode (TSM) or quartz crystal microbalance (QMB or QCM) sensors.

Gravimetric odor sensors using acoustic wave devices which are operated by detecting the effect of adsorbed molecules on the propagation of acoustic wave have also been investigated for application to an E-nose [16]. Figure 6.10 shows one type of gravimetric odor sensor used in E-nose system. Table 6.3 describes various types of E-nose sensors and material used in manufacturing of the sensor and operating principle of the same.

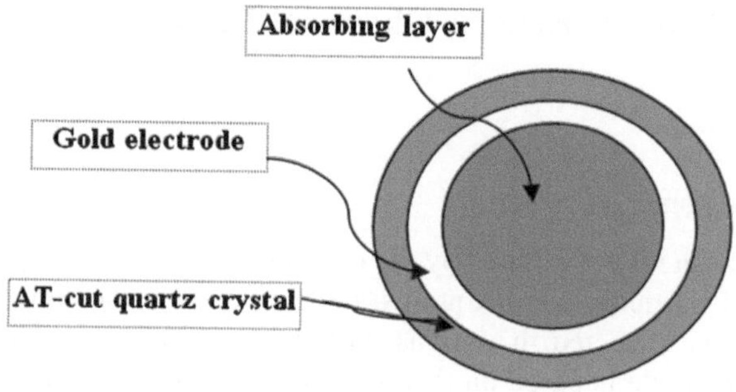

Fig. 6.10 Gravimetric odor sensor

Table 6.3 Types of E-nose sensor and material usage and detection principle of sensors [18]

Sensor type	Sensitive material	Detection principle
Acoustic sensors: Quartz crystal microbalance (QMB); surface and bulk acoustic wave (SAW, BAW)	Organic or inorganic film layers	Mass change (frequency shift)
Calorimetric; catalytic bead (CB)	Pellistor	Temperature or heat change (chemical reaction)
Catalytic field-effect sensors (MOSFET)	Catalytic metals	Electric field change
Colorimetric sensors	Organic dyes	Color changes, absorbance
Electrochemical sensors	Solid or liquid electrolytes	Current or voltage change
Fluorescence sensors	Fluorescence sensitive Detector	Fluorescent-light emissions
Infrared sensors	IR-sensitive detector	Infrared-Radiation Absorption
Metal oxides semi-conducting (MOS)	Doped Semi-conducting metal oxides (SnO_2, GaO)	Resistance change
Optical sensors	Photodiode, light sensitive	Light modulation, optical changes
MPEN	Micro plasma	Light emission, spectra analysis

Two types of acoustic wave E-nose sensors are mostly used and discussed here:

(a) Quartz crystal microbalance (QCM) sensor also known as bulk acoustic wave (BAW),
(b) Surface acoustic wave (SAW).

6.7 Chemo Sensor (Electro-Chemical Sensors)

- Chemoresistor
- Chemocapacitor
- Chemotransistors

A chemical state is determined by the different concentrations, partial pressures, or activities of particles such as atoms, molecules, ions, or biologically relevant compounds to be detected in the gas, liquid, or solid phase. The chemical gas sensor basic principle is based on that analyte molecules come into contact with a chemically sensitive material which cause a change in the properties of the material, follow-on in a change in the electrical signal. The most common material used for gas sensors is the tin-dioxide semiconductor, which is doped in order to provide selectivity.

Typically, the sensors are contained in an instrument that regulates the flow of air. Sampling is done in three phases such as baseline, sampling, and recovery phases. The gas to be analyzed is exposed to the sensor array in the sampling phase while a reference gas is used during baseline and recovery phases in order to return the sensor values to the initial state. Valuable information is not only obtained in the transient and static sensor values in the sampling phase but also obtained in the dynamic appearance in the recovery phase.

6.7.1 Chemoresistor Sensors

The typical structure of a planar chemoresistor is shown in Fig. 6.11 that employs constant voltage circuit. The resistance of the chemically sensitive layer changes with the amount of adsorbed chemical. The resistivity measurement is done either via a Wheatstone bridge arrangement or by recording the current at an applied voltage in a direct current (DC) mode or in a low-amplitude, low-frequency alternating current (AC) mode to avoid electrode polarization.

1. Metal Oxide Semiconductor (MOS)

MOS sensors are used for observing the electrical-resistance changes that occur when vapors are adsorbed onto a semiconductor surface [17]. MOS sensor contains a ceramic structure heated by wire and coated with a metal oxide semiconducting film. So usually it is called metal oxide or ceramic gas sensors [18]. MOS sensor also label as Taguchi and Figaro because of the inventor and the first company to manufacture likewise.

In general, sensors are prepared by depositing a thin porous film of a metal-oxide material (usually tin oxide) onto an electrically heated ceramic pellet and annealing at high temperatures. The oxygen available in the air adsorbed onto the sensor surface removes electrons from the conduction band of the semiconductor

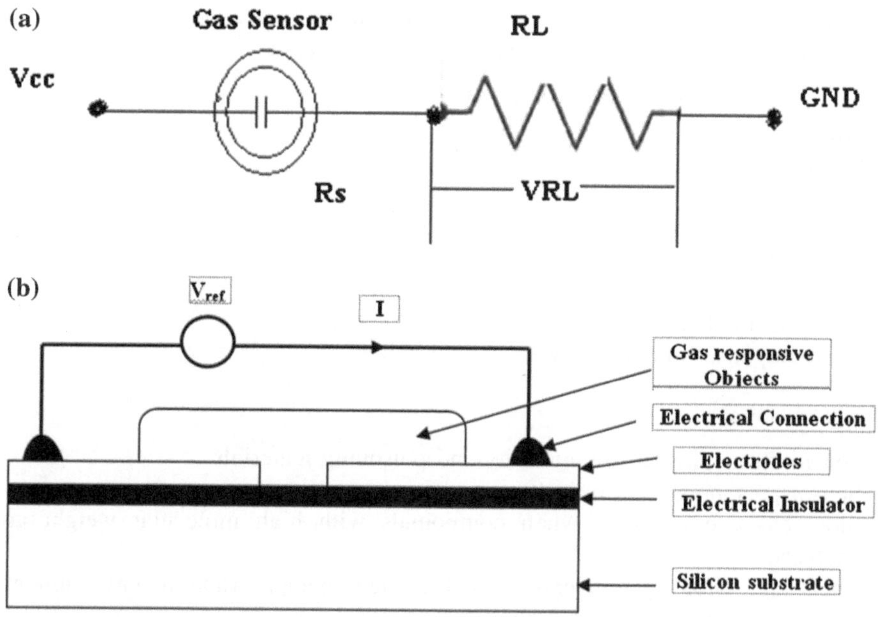

Fig. 6.11 **a** Representation of gas sensor, **b** structure of chemoresistive gas sensor system

and increases electrical resistance of sensor. Electron trapping is decreased by interface of gases with the surface adsorbed oxygen, leading to increases in electrical conductance of the sensor. Consecutively to reduce response and recovery times, metal-oxide sensors are typically run at elevated temperatures (up to 400 °C).

Devices that translate the changes in the concentration of gaseous chemical type into electrical signals are distinct as metal oxide sensors. MOS sensor basic schematics as shown in Fig. 6.12 consist a sensitive layer, an insulating layer, two electrodes, and a heating layer. At temperature level of 250–450 °C, the semiconducting layer oxidizes the sample compound. Therefore, the semiconducting

Fig. 6.12 Basic metal oxide sensor

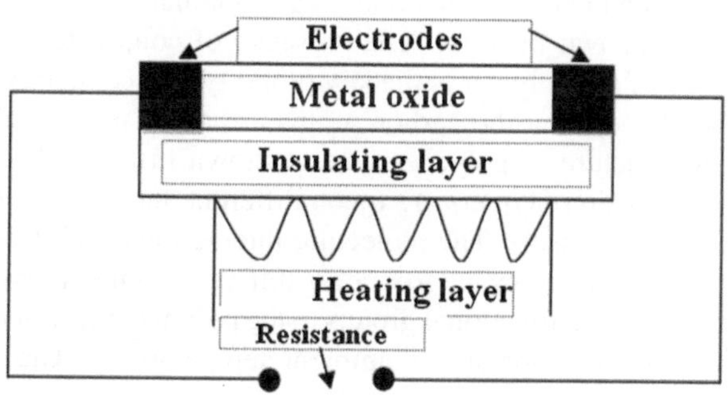

substance adsorbs the free electrons and its conductivity changes. As a result, the change of resistance in the electrical circuit is registered.

Advantages:

- Small size sensors.
- Robust.
- Quite good sensitivity.
- Very low cost.
- Electronic simplicity.
- Long-lasting life to operate.
- Easily fabricated.
- Short response time to analytes.

Disadvantages:

- Sensitivity is affected by humidity and poisoning materials.
- Sensor poisoning, e.g., by sulfur compounds, etc.
- Slow baseline recovery when compounds with high molecular weight are analyzed.
- Their high-working temperature makes them inappropriate in environments containing flammable chemicals.
- Poor specificity and selectivity.
- Sensor drift mainly caused by sensitivity loss.
- Signal analysis is difficult due to sensor signal is generally not linear.
- High power level is needed to run the sensors.

Application:

- Detection of gas leakage.
- Detection of combustible and toxic gases.
- Detection of VOC of different foodstuffs such as coffee, milk, strawberry, juice, meat, etc.

2. Conducting Polymers

Polymers (plastics) are identified as good insulating properties and most used materials in the modern world. CP are polymer materials with metallic and semiconductor characteristics, a combination of properties not exhibited by any other known material. The presence of conjugated double bonds along the backbone of the polymer is the key property of a conductive polymer. Also note that the bonds between the carbon atoms are alternately single and double. For example, the structure of polyacetylene is shown in Fig. 6.13.

Organic polymers are much different and can report a wide variety of functionalities to sensors. The molecular interaction capabilities of a polymer can be selectively modified by incorporating different counter ions during polymer preparation or by attaching functional groups to the polymer backbone. A different advantage of CP is that they operate at ambient temperatures. The sensors can detect odors at

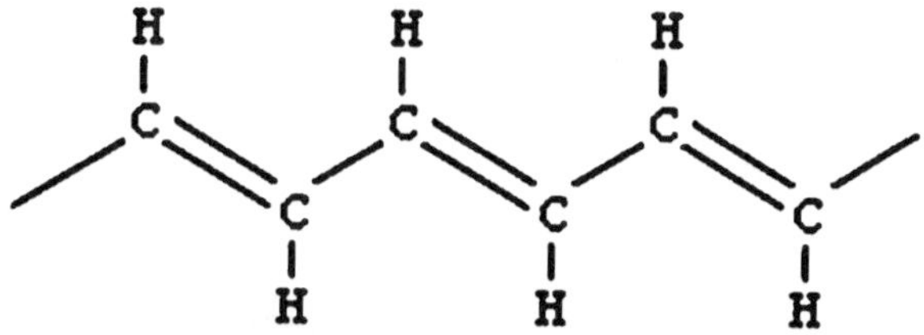

Fig. 6.13 Polyacetylene chain

sensitivities of 0.1 ppm (ppm), but 10–100 ppm is more usual. In general, available CPs are polyacetylenes, polypyrroles, polythiophenes, polyterthiophenes, polyanilines, ployfluorines, poly-3-alkylthiophenes, polytetrathiafulvalenes, polynapthalenes, polyphenylene sulfide, poly-phenylenevinylenes, poly-3,4-ethylenedioxythiophene, polyparaphenylene, polyazulene, polyparaphenylene sulfide, polycarbazole, and polydiaminonaphthalene.

The active materials used are the polypyrroles, thiophenes, indoles, or furans (shown in Fig. 6.14). The conductivity of these materials changed as they are exposed to various types of chemicals, which bond with the polymer backbone. This bonding may be ionic or in various cases covalent. The interaction affects the transfer of electrons along the polymer chain, that is to say, its conductivity. A given compound's affinity for a polymer and its effects on the polymer's conductivity are strongly influenced by the counter-ions and functional groups attached to the polymer backbone. The use of conducting polymer nano composites/nano particles could greatly improve diffusion since they have much greater exposed surface area and as a result of this the basic characteristics of a biosensor like low detection limit get enhanced.

Conducting materials change resistance when they come in environment having chemicals/gases. As soon as the polymer composite sensor is exposed to a vapor of chemical/gases, some of the vapor permeates into the polymer and causes the polymer film to expand. The vapor-induced expansion of the polymer composite causes an increase in the electrical resistance of the polymer composite because the polymer expansion reduces the number of conducting pathways for charge carriers [19] (Fig. 6.15).

Advantages:

- It can be prepared both electrochemically and chemically.
- It can be prepared in a range of soluble and insoluble forms.
- It has unique electrical, electronic, magnetic, and optical properties.
- Compliance with micro and nanoscale fabrication.
- Compatibility with diverse range of fabrication techniques such as electrochemical, optical, mass-based, etc.
- Biomaterials such as enzymes, antibodies, whole cells, and nucleic acids can be incorporated into the polymer matrix.
- Strong biomolecular interactions.
- Low detection limits.
- Enhanced sensitivity (when used as a composite material with nanoparticles).

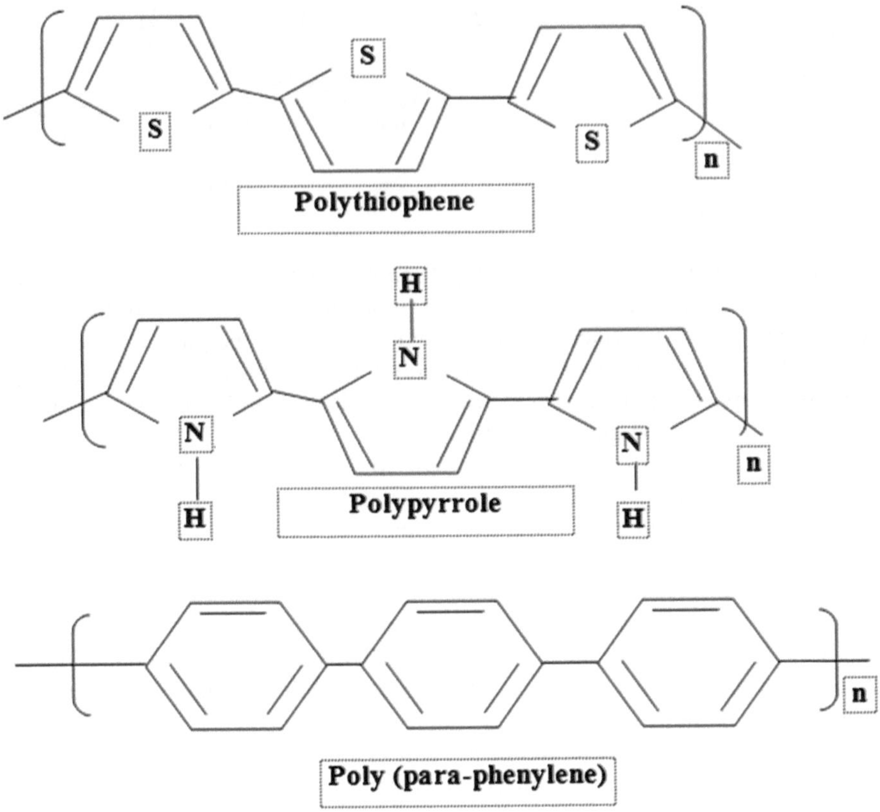

Fig. 6.14 Active materials used in CP sensor

- Reversible responses at ambient temperatures.
- Cost-effectiveness.

Disadvantage:

- Long response times (20–40 s) or a rapid drift of their responses over time.
- Inherent time- and temperature-dependent drift.
- Poor batch-to-batch reproducibility.
- The high cost of sensor fabrication.
- Greater sensitivity than metal oxides to water vapor renders them susceptible to humidity.
- Extremely sensitivity to moisture due to their low operating temperature (<50 °C).

Application:

- Detect alcohols, acetone, benzene, and other polar volatile chemicals.

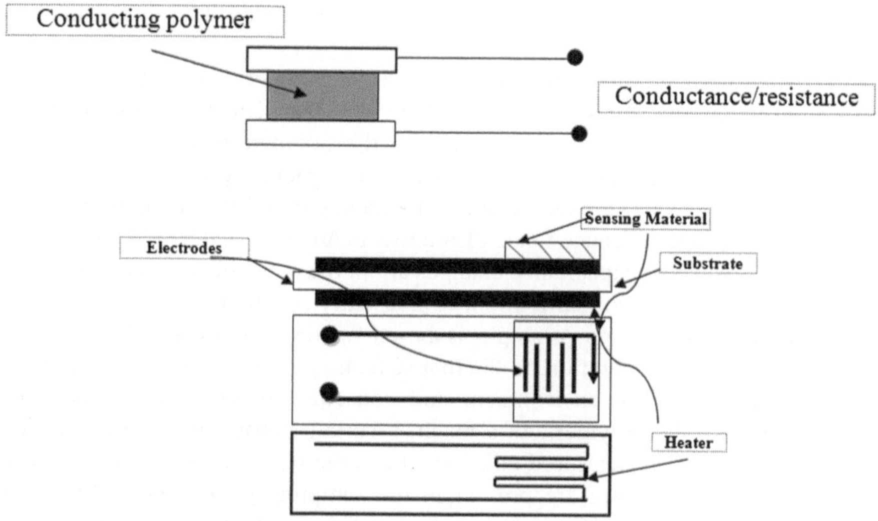

Fig. 6.15 Schematic of conducting polymer [28]

- In foodstuffs like beer and other alcoholic beverages.
- Olive oil, orange juices, fishes, and milk.
- Virgin olive oils characterized by different levels of oxidation.

6.7.2 Chemocapacitive Sensor

Capacitive sensors sometimes identified as chemocapacitors are category of electrochemical sensor. The capacitive sensor consists of two sets of interdigitated electrode structures, which correspond to the two plates of a standard capacitor (Fig. 6.16). The sensor monitors changes in the dielectric coefficient of the polymer between electrodes upon analyte absorption. Therefore, chemocapacitors

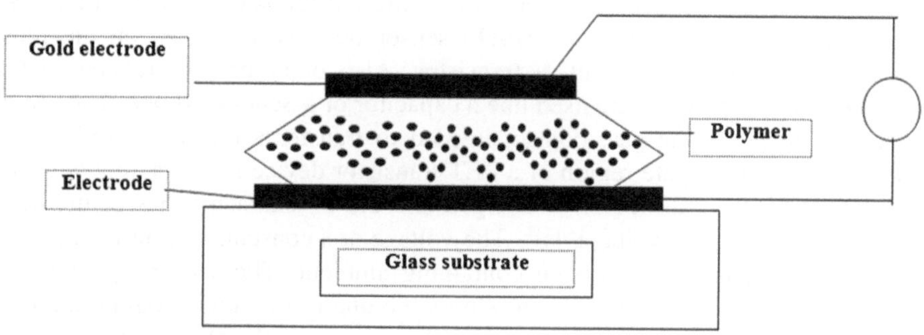

Fig. 6.16 Chemo capacitor representation

(di-electrometers) rely on changes in the dielectric properties of a sensing, material upon analyte exposure.

Two effects change the capacitance of, e.g., a polymeric sensitive layer upon absorption of an analyte: swelling and change of the dielectric constant due to incorporation of the analyte molecules into the polymer matrix. Interdigitated electrode structures are predominantly used for capacitance measurements. The sensor devices generally are operated at a frequency of a few kHz up to 500 kHz. Because of the nominal capacitance of microstructured capacitors (of the order of 1 pF) and the expected capacitance changes are in the range of some attoFarads, an integrated solution with on-chip circuitry is usually required.

Chemocapacitor sensors principle is based on the two steady states for the sensitive layer during operation. In the first state, no gaseous analyte molecules are present in the sampling environment and consequently only air is, therefore, incorporated into the polymer. As a result, a certain capacitance (C) of the sensitive polymer layer is measured and constitutes the baseline. In the second state, gaseous analyte molecules are present in the sampling environment. When the polymer absorb the gaseous analyte, the sensitive polymer layer changes its electrical (e.g., dielectric constant) and physical properties (e.g., volume V) to produce deviations (De, DV) from the first state (reference state). The changes in electrical and physical properties of polymers are the result of reversible incorporation of gaseous analyte molecules into the polymer matrix.

Although capacitive sensor responds to capacitance changes, the output signal is converted to a frequency and this change is delivered as a differential signal generated using a Sigma-Delta-modulator circuitry between a passive reference and a polymer-coated sensing capacitor.

6.7.3 Chemotransistor

Field-effect-based transistors, which are the most common electronic components on current IC logic chips, rely on modulation of the charge carrier density in the semiconductor surface space-charge region through an electric field upright to the device surface: an isolated gate-electrode controlled the source-drain current. So briefly, the field effect transistor (FET) sensor devices are based on the most common electronic component, a transistor. Also other devices related to the transistor but more simple are used like a capacitor or a Schottky diode. The use of a catalytic metal as the gate contact is used to transfer a transistor device to a sensor. The catalytic gate region of a FET transistor device is interacting with the gases. Thereby the gate region is charged and the IV characteristics of the transistor change along the voltage axis. The voltage at a constant current is higher in oxygen as compared to hydrogen containing ambient. The sensor signal is the voltage at a constant current, which is possible due to the diode configuration of the transistor (gate and drain or gate and source are short). In this way, the

Fig. 6.17 Representation of FET

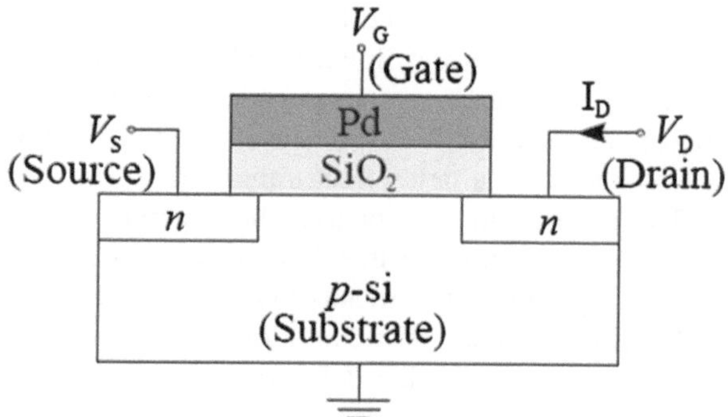

time-dependent signal can be monitored and the response is easy to calculate. Basic structure representation of FET is shown in Fig. 6.17.

The source-drain conductivity of the FET can be modulated by adjusting the strength of electrical field between the gate electrode and the silicon, or through the presence of absorbed species and the resulting polarization phenomena. In FET, the influence of sample gases change the value of the transistor threshold voltage.

6.8 Metal Oxide Semiconductor Field Effect Transistor

The Metal Oxide Semiconductor Field Effect Transistor (MOSFET) is the leading device used and has taken the limelight compare to the other technologies for a number of reasons, though mainly due to its simple structure. It is a transducer device used in E-nose to transform a physical/chemical change into an electrical signal. It is basically a four-terminal device (as shown in Fig. 6.18) in which current flow is controlled by an externally applied vertical electric field. When

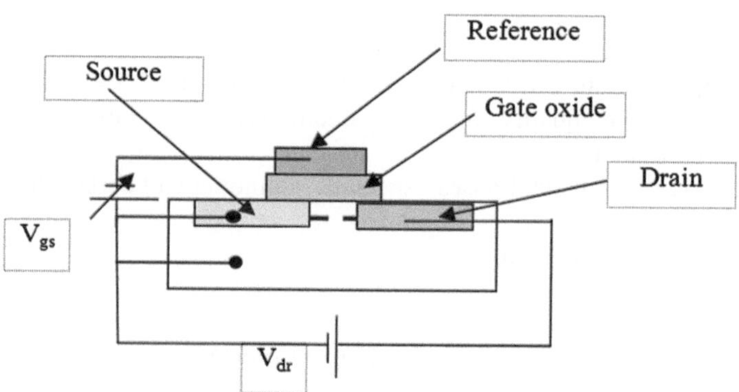

Fig. 6.18 MOSFET schematic

voltage is not applied, the back-to-back p-n junction between the drain and the source prevents current flowing. When a positive voltage applied on the gate in respect to the source, negative carriers provide a conduction channel between the source and the drain. Vertical as well as the lateral field controlled the current so it is also known as a field effect transistor.

The microchemo sensor uses the structure of a MOSFET in which the gate is made of a gas sensitive metal such as Pd as it was first proposed by Lundstrom et al. in 1975 [20]. The metals that compose the gates of a transistor are replaced with catalytic metals or metal alloys (e.g., platinum (Pt), palladium (Pd), iridium (Ir), alloys, etc.) and then left exposed to air. The interaction of adsorbed gases alters the surface-charge density and thus changes the potential of the device. The choice of the operational temperature, the metal used on the gate and varying the microstructure of the metal help in achieving the MOSFET selectivity.

The general structure of the MOSFET is shown in Fig. 6.19. Three components contained in a sensor are: a catalytic metal, an insulator, and a transistor (semiconductor). The device basic principle is to allow gaseous compounds to react with the catalytic metal and produce species that are able to diffuse through the metal film and absorb onto a metal insulator. Due to the absorption voltage will change and it will change the current voltage characteristics of the sensor (as shown in Fig. 6.19). Different catalytic metals may be used to vary the sensitivities of the sensor. When VOCs come in contact with a catalytic metal, it produces a reaction, the products which diffuse through the gate changing its surface potential. The voltage shift depends on the gas concentration. The sensitivity and selectivity can be enhanced by varying the thickness/type of metal catalyst and changing the operating temperature, usually 100–200 °C. They too are susceptible to drift similar to conductivity sensors.

Advantage:

- MOSFET are robust sensors.
- Low sensitivity to humidity.
- Sensor reproducibility is good.
- Low-cost sensor.

Disadvantage:

- Running temperature can affect the selectivity and sensitivity of sensor.
- To achieve good quality and reproducibility, high-quality manufacturing expertise is needed.
- Low sensitivity to defined gases especially ammonia and carbon dioxide.
- Baseline drift.
- Low sensitivity to moisture.

Application:

- Food cooking, fermenting, and wine-making processes.
- Ethylene measurement during fruit ripening process.

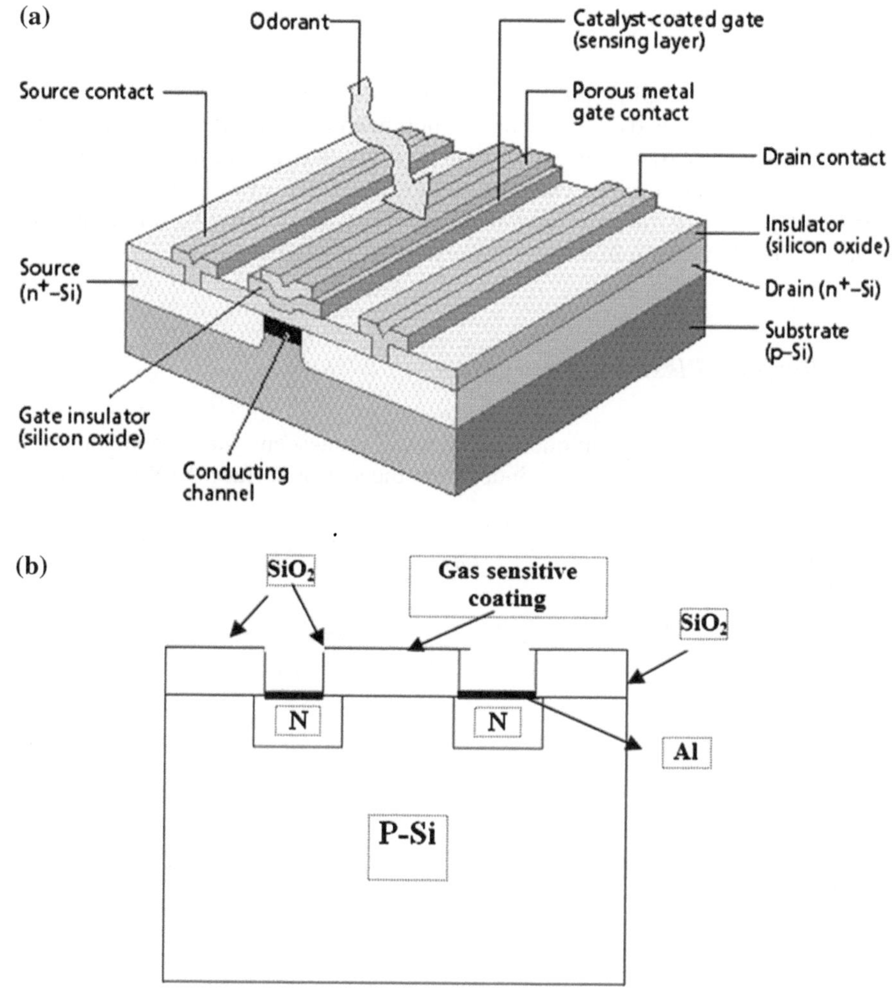

Fig. 6.19 **a** MOSFET structure [29], **b** MOSFET gas sensor (Sio_2 deposited on *top*)

6.9 Optical Fiber

Optical sensor systems are relying on changes in fluorescence, wavelength absorption, or reflection. Like other kinds of biosensors and E-noses in particular, these sensors make the most of the interaction between molecules in the gaseous phase, which are adsorbed onto a sensitive layer and result in changes of their optical properties.

6.9.1 Optical Fiber

Optical fiber sensors utilize glass fibers coated with thin chemically active materials on their sides or ends that contain immobilized fluorescent dyes in an organic polymer matrix as shown in Fig. 6.20. The changes in dye polarity cause a shift in the emission spectrum (color change) due to interaction of volatiles with a light source. Wide sensitivities are obtainable owing to availability of different dyes; however, they have a limited lifetime because of photobleaching.

6.9.2 Surface Plasmon Resonance

The SPR is an optical technique that uses the evanescent wave phenomenon to measure changes in the refractive index near the sensor surface. SPR is an optical

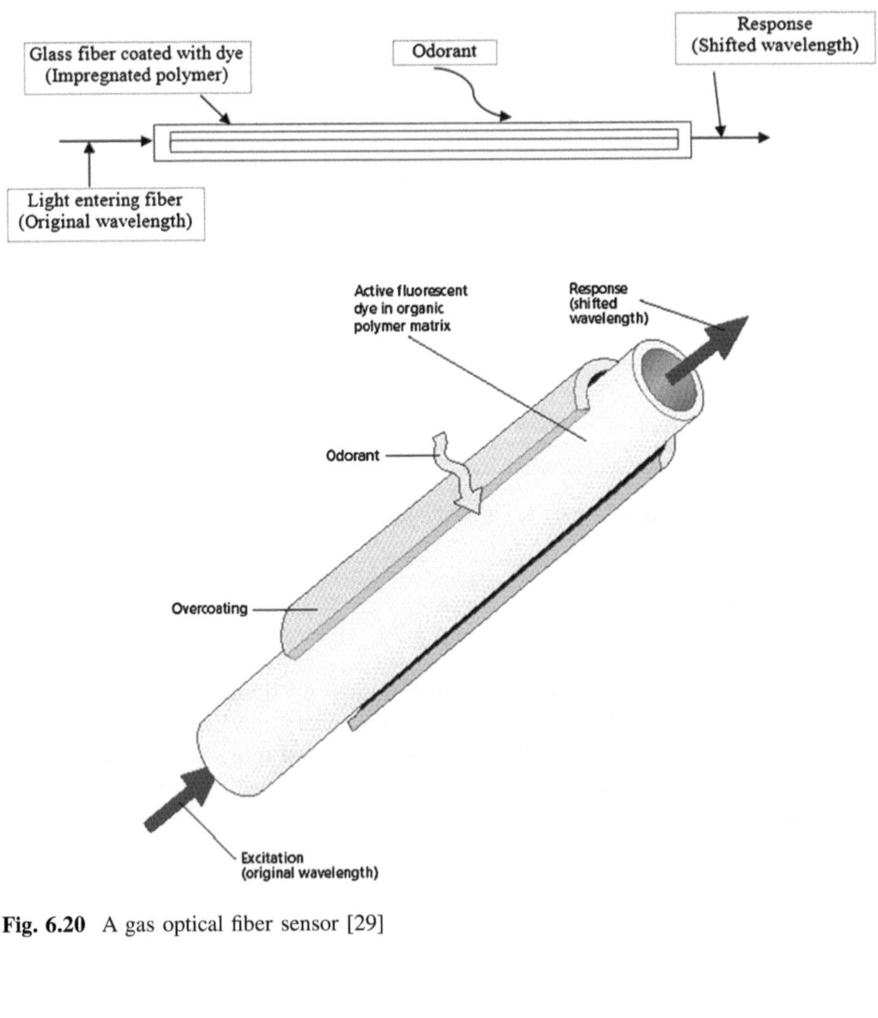

Fig. 6.20 A gas optical fiber sensor [29]

phenomenon in which incident light excites a charge-density wave at the interface between a highly conductive metal and a dielectric material. The conditions for excitation are determined by the permittivity of the metal and the dielectric material. For measuring small changes in the refractive index of a thing region adjacent to the metal surface, the transduction principle of SPR is basically used as an analytical tool. For sensitive detection of chemical species such as odor, vapor, and liquid, the optical excitation of surface plasmon on a thin metallic film has been recognized as a promising technique. To monitor the excitation of SPR by measuring the light reflected from the sensor interface, several methods have been employed which include analysis of angle modulation, wavelength modulation, intensity modulation, and phase modulation. Optical SPR sensors are sensitive to change in the refractive index of a sample surface. The change in refractive index of a sample surface sensitivity is measured by optical SPR sensors. Gases such as ammonia, toluene, xylene, ethylacetate, 4-methyl-2-pentanone, and propionic acid can be detected by measuring the SPR using angle modulation. The SPR is measured with a prism and a thin highly conductive gold metal layer deposited on the prism base. To excite the SPR, the LED emitting 660 nm light is used as light source.

The reflection spectrum (reflected light intensity versus angle of incidence with respect to the normal of metal/dielectric interface) of SPR is measured by coupling transverse magnetically polarized monochromatic light into the prism and measuring the reflected light intensity of the ray exciting the prism versus the incidence angle. A very thin film of methyl-methacrylate, polyester-resin, or propylene-ether as a sensing membrane is deposited on gold metal thin film using spin-coating method to utilize this system as a gas sensor. Using CCD camera attached to a personal computer the reflected light is measured. The angle at which the minimum reflection intensity occurs is the resonance angle at where coupling of energy occurs between the incident light and the surface plasmon waves. Four channel images of reflected light are observed by using the CCD camera. The SPR sensor with synthetic polymer thin film on the gold metal film as a sensing membrane exhibits high sensitivity for toxic gases such as ammonia, toluene, xylene, ethyl acetate, 4-methyl-2-pentanone, and prop ionic acid.

6.9.3 Fluorescent

Another sensing device that is designed as an array of optically based chemosensors providing input to a pattern recognition system on the E-nose technology has been developed. This type of chemosensor consists of optical fibers deposited with fluorescent indicator Nile Red dye in polymer matrices of varying polarity, hydrophobicity, pore size, elasticity, and swelling tendency to create unique sensing regions that interact differently with vapor molecules [15].

Fiber-optic sensors most often consist of an analyte sensing element deposited at the end of an optical fiber. Individual optical fibers with a diameter as small as

2 μm and imaging bundles with a diameter of 500 μm are available. In the fiber-optic chemosensing system, the optical sensing element is typically composed of a reagent phase immobilized at the fiber tip by either physical entrapment or chemical binding.

This reagent phase usually contains a chemical indicator that experiences some change in optical properties, such as intensity change, spectrum change, lifetime change, and wavelength shift in fluorescence upon interaction with analyte gases or vapors. The responses depend upon the nature of the organic vapor and the strength of its interaction with different polymer systems used.

Advantage:

- Excessive sensitivity.
- Identifications of individual compounds in mixtures.
- Capabilities of finding multiparameter.

Disadvantages:

- Connected electronics and software are very complex.
- Connected electronics and software also leading to increased cost.
- The sensors have quite a short lifetime (due to photobleaching).
- Complex sensor-array systems.
- Low portability due to delicate optics and electrical components.

Application:

- Fire detector.
- Simple gases detect such as O_2, CO_2, CH_4, HCL, and HF.
- Detection of pathogens.
- Medical diagnosis based on protein or cell concentration.
- Real-time detection of DNA hybridization.

6.10 QCM/QMB

QCM is an advanced type of microbalance mass sensor. The transducer for the QCM sensor is mass-sensitive and very similar to the SAW sensor. Only distinct between SAW and QCM is that the sensor employment. SAW uses a surface acoustic wave sensor while QCM uses a BAW sensor. The device (electro-acoustic) based on the Piezo-electrical properties of quartz material has been implemented in sensors.

QCM oscillation frequency decreases while the mass is bound on the crystal surface, therefore it is identified as very sensitive mass measuring device. QCM sensing mechanism is based on the shift in the quartz crystal (QC) resonant frequency due to the adsorption of gas molecules onto the sensing films. QC which is equipped with metal electrodes (e.g., gold) is the basic material of the QCM

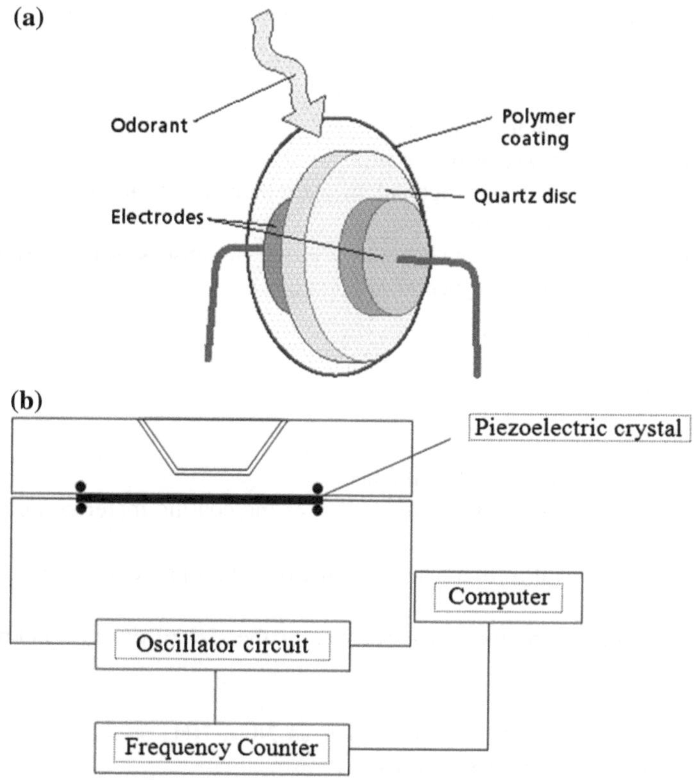

Fig. 6.21 **a** The quartz crystal microbalance (QCM) sensor [29], **b** scheme of a flow cell for piezoelectric crystals (the oscillator circuit and frequency counter)

sensor. It is also known as a BAW device, which is made of a polymer-coated resonating quartz disc, vibrating at a characteristic frequency (10–30 MHz).

As the volatile molecules adsorbed to the polymer surface, it increases mass of the disc, and thus it reduces the resonance frequency. The decrease is inversely proportional to mass of odorant adsorbed. Changes in temperature, humidity, and flow conditions can affect sensor sensitivity. The sensor selectivity is dictated by the thickness of the coatings. Sensor surface is coated by a sensitive material, which is used to enable detection of the measurand in the environment. For conversion of the measurand quantity to an electrical signal, an appropriate electronic circuit is required. QCM sensor is a very good sensitive detector of mass changes.

The basic working principle of the QCM sensor is represented in Fig. 6.21. In the surrounding space of a QCM sensor, analytes are present and it is interacting with the sensitive coating material on the sensor surface. During the interaction, analyte molecules are absorbed into the sensitive coating material, i.e., polymer. The mass change on the sensor surface is occurred due to the adsorption of the analytes by the coating materials. Thereafter the mass change on the sensor surface is converted to the frequency change.

Advantages:

- For large dynamic range its linear.
- High sensitivity and fast response time.
- Can be implemented in a variety of measurement locations and various situations as its stabile long term.
- It can be used as a single sensor or as a group of sensors (sensor array).
- Detection of trace gases is special application.
- Low power utilization.
- Miniature size.
- Robustness.

Disadvantages:

- Temperature and humidity dependence.
- The poor reproducibility in the deposition of the coating material, i.e., complex fabrication process.
- Need of a thin crystal to increase the sensitivity that prevents their inclusion in multisensory arrays.
- Poor signal-to-noise performance because of the size of crystal thickness.

Application:

- Used to detect gases and volatiles such as NO_2, NH_3, xylene, toluene, tetrachloroethylene, H_2, and toxic gases.
- Organophosphorous compounds, SO_2.
- To analyze the headspace of foodstuffs as olive oil samples.

6.11 SAW

The sensors having their design based on surface acoustic wave (SAW) platform are perhaps the most sensitive, miniature of E-nose gravimetric type of sensors. Oscillator circuits whose resonance frequencies are controlled by resonator devices in the feedback path are defined as SAW sensors. Chemical vapor sorption by depositing broadly selective polymer thin films in the acoustic wave propagation region is the functional principle of the SAW devices. An oscillator frequency is shifting due to exposure to vapor and it is taken to the chemical signal [21]. The schematic of SAW sensor is shown in Fig. 6.22.

SAW sensors require waves to travel over the surface of the device. SAW sensors operate at higher frequencies (100–1,000 MHz) and thus generate a larger change in frequency. SAW sensors also suffer from reduced long-term stability and are highly sensitive to humidity. Main characteristics of SAW sensors are its

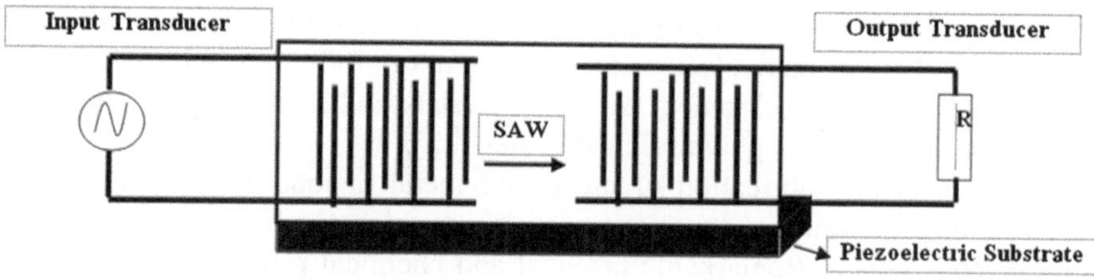

Fig. 6.22 SAW sensor

continuous upgradability in performance through increase in operation frequency, variation in device design, development in polymer interfaces, and planner technology.

Advantages:

- High sensitivity and fast response times.
- Fabrication is easy.
- Diverse sensor coatings.
- Small.
- Inexpensive.
- Sensitive to virtually all gases.

Disadvantages:

- Poor signal-to-noise performance because of the requirement of high frequencies to operate.
- Required circuit for operation is expensive and complex.
- Reproducibility is difficult to achieve.
- Complex circuitry.
- Temperature and humidity sensitive.
- Specificity to analyte groups affected by polymeric-film sensor coating.

Applications:

- Analysis of dairy products, perfumes, and alcoholic beverages
- Environmental analysis.
- Medical diagnosis.

6.12 Advanced Sensors

Multisensor systems implemented in E-nose are the advanced sensing devices designed to analyze complex analytes mixtures. In the last few years, new challenges have emerged in sensor market, in particular a growing demand for chemical sensors adaptable to nonplanar surfaces, weightless, inexpensive, and with integrated front-end electronics. Hence, new low temperature fabrication technologies like printed electronics or system on foil electronics are deeply investigated for the development of new materials and manufacturing processes compatible with flexible polymeric substrates [22].

Advances in the fabrication of metal and metal nanoparticles have deferred nanostructured materials with unique properties, which can be potentially applied to the E-nose sensor. Remarkable physical and chemical properties of metal and semiconductor nanoparticles also provide them as promising materials in the fields vary from optoelectronics sensor to medicine sensor.

Zampetti et al. [23] show the design, fabrication, and test of a fully flexible sensorial system (SS), composed of three dissimilar sensor units (SU), integrated on an ultrathin polyimide substrate (PI) of 8 mm thick. Each SU consists of a capacitive sensor connected to a ring oscillator (RO) used as readout circuit. In each unit, they utilized a different chemical interactive material (CIM) as dielectric polymer: poly(tetrafluoroethene) (PTFE), poly(methyl 2-methylpropenoate) (PMMA) and benzocyclobutene(BCB). These materials can be easily handled since they are commonly used in microelectronic process as passivation layer or photo-resist. These characteristics make possible a large scale production.

Similarly Kinkeldei et al. [24] works with the chemiresistors type E-nose sensor and demonstrated a method to integrate an E-nose system into a textile band to fabricate a smart textile. The demonstrated sensor system consists of four different polymer composite gas sensors on a flexible polymer foil. With this E-nose inside a textile, it could differentiate between the exposures of four solvents using the solubility parameter concept.

Giménez et al. [25] introduce PAMPA III E-nose, which is an equipment operating with a measurement system based on a set of 12 semi-conductive Metallic Oxide Gas Sensors. Thin film Micro-Electro-Mechanical Systems (MEMS) type sensors are built by evaporation deposition of thin semiconductors (SnO_2, ZnO, or WO_3, among others) films, which must be heated locally in a temperature range from 300 to 450 °C to detect the different gaseous species. Metallic oxide semiconductors are usually doped with different elements (Al, In, Pd, Au, Sb, etc.) to increase sensor selectivity.

Development of chemical sensors (electrochemical biosensors) consisting of CP nanocomposite materials produced by the electropolymerization of CPs on to specialized nanoparticle electrodes.

ST&D developed the microplasma-based E-nose (MPEN) which is an innovative approach within the area of gas sensing devices and provides a very flexible device, which can detect almost any chemical without prior knowledge or calibration.

Spinelli et al. [26] evaluated the use of a near infrared (NIR) instrument in combination with an E-nose system for the early detection of fire blight (disease) in pears. The E-nose system detected the disease prior to symptom development by the distinctive olfactory signature of volatiles released as early as 6 days after infection.

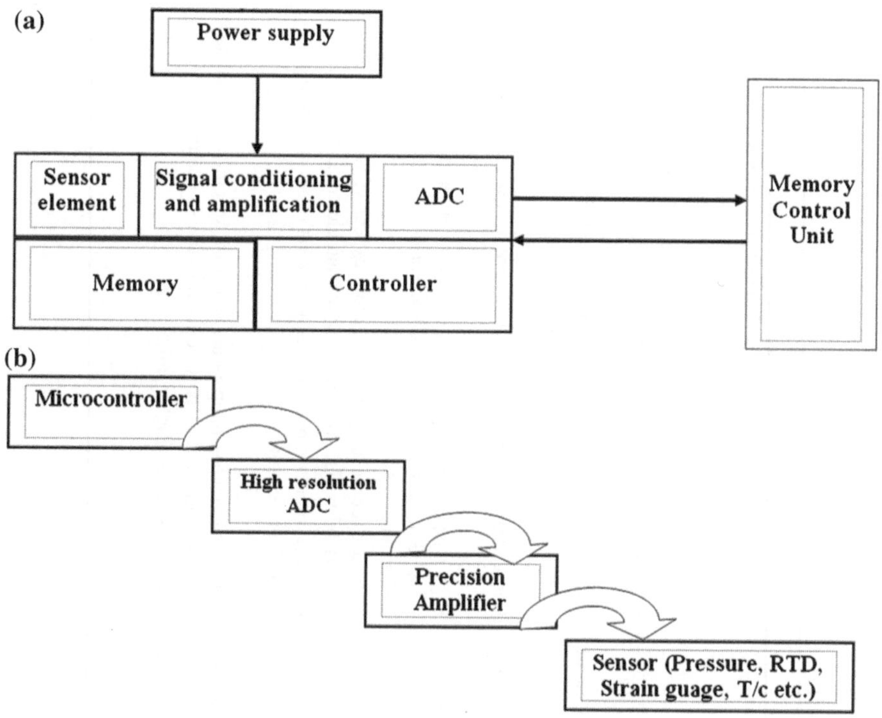

Fig. 6.23 **a** Smart sensor system (*schematic block*), **b** basic elements of smart sensor

Sankaran et al. [27] reviewed other advanced techniques and instruments for detecting plant diseases, which might be used in combination with E-noses for disease diagnoses.

6.12.1 Sensory Array

Sensor arrays contain a plurality of discrete sensor elements each of which provides information about single- or multi-modal stimuli. The array composition and structure can be extensive in quantity and dimension. In biomedical applications, several medical diagnostic techniques also take advantage of sensor arrays. The system is more flexible in use compared to the individual parts and thereby appropriate for more applications. Artificial electronic tongues and noses have generally employed arrays of sensors. In such systems, arrays of microcantilevers are used to detect the presence of specific substances based on (electro-) chemical reactions, which can be used for automatic classification and recognition [19]. High discrimination in array sensors can be easily achieved using conducting polymer materials due to the wide range of polymeric materials available on the market.

Table 6.4 Summary of advantages and disadvantages of E-nose sensor types [18]

Sensor type	Advantages	Disadvantages
Calorimetric or catalytic bead (CB)	Fast response and recovery time, high specificity for oxidized compounds	High temperature operation, only sensitive to oxygen containing compounds
Catalytic field effect Sensors (MOSFET)	Small sensor size, inexpensive operating costs	Requires environmental control, baseline drift, low sensitivity to ammonia and carbon dioxide
Conducting polymer sensors	Ambient temperature operation, sensitive to many VOCs, short response time, diverse sensor coatings, inexpensive, resistance to sensor poisoning	Sensitive to humidity and temperature, sensors can be overloaded by certain analytes, sensor life is limited
Electrochemical sensors (EC)	Ambient temperature operation, low power consumption, very sensitive to diverse VOCs	Bulky size, limited sensitivity to simple or low mol. wt. gases
Metal oxides semiconducting (MOS)	Very high sensitivity, limited sensing range, rapid response and recovery times for low mol. wt. compounds (not high)	High temperature operation, high power consumption, sulfur and weak acid poisoning, limited sensor coatings, sensitive to humidity, poor precision
Optical sensors	Very high sensitivity, capable of identifications of individual compounds in mixtures, multiparameter detection capabilities	Complex sensor array systems, more expensive to operate, low portability due to delicate optics and electrical components
Quartz crystal microbalance (QMB)	Good precision, diverse range of sensor coatings, high sensitivity	Complex circuitry, poor signal-to-noise ratio, sensitive to humidity and temperature
Surface acoustic wave (SAW)	High sensitivity, good Response time, diverse sensor coatings, small, inexpensive, sensitive to virtually all gases	Complex circuitry, temperature sensitive, specificity to analyte groups affected by polymeric- film sensor coating
MPEN	One for all, no application specific design, fast response, excellent sensitivity, higher stability-small size, disposable, good accuracy	Complex system, Higher initial cost

Table 6.5 Available electronic nose instrument [18, 28]

Manufacturer name	Application	Website	Sensory types	Model name
AIRSENSE Analysentechnik GmbH	Food evaluation; flavor and fragrance testing, food stuff control, chemical analysis, medicines, environment safety	www.airsense.com	MOS sensory array	PEN Portable Electronic Nose PEN2 I-PEN i-PEN ET i-PEM PCN iPEN3/MOD
Alpha MOS	Analysis of food types; quality control of food storage, fresh fish, and petrochemical products; packaging evaluation; analysis of dairy products, alcoholic beverages and perfumes	www.alpha-mos.com	QMB/CP, MOS	Fox GEMINI
GSG Analytical Instruments.	Classification of gas mixtures, aromas and flavors	www.gsg-analytical.com	Multiple sensor options including QCM, MOS, QMB	M·O·S·E·S II VOCvario VOCcheck VOC meter VOC iDent
Seacost science	Specific classes of chemicals	www.seacoastscience.com	CP	SC-210 SC-M 100
MIL-RAM Technology, Inc.	Detect hazardous gases, toxic and combustible gases for environmental protection and process controls, as well as oxygen deficiency for personal safety.	www.wirelessmill-ram.com www.mil-ram.com	CP Electrochemical	Tox-array 2100
Electronic sensor technology	Detect and analyze accurately organic, biological & chemical compounds in real time.	www.estcal.com	SPR	ZNOSE (portable, battery operated, benchtop)

(continued)

Table 6.5 (continued)

Manufacturer name	Application	Website	Sensory types	Model name
Environics	Gas and oil production and pipeline safety, plastics Manufacturing, semiconductor fabrication, chemical processing, medical sterilization, solar and flat panel production, DUV photo resist development, high purity filter manufacturing, pharmaceutical production and sterilization, fluoropolymer contamination	http://www.environics.fi	Photoionization detector	ENVI-VOC
Applied sensor	Indoor air quality monitor	www.appliedsensor.com	MOS, MOSFET,	Indoor air monitor
RaeSystemes	Portable handheld VOC monitor extended range of 0–15,000 ppm makes it an ideal instrument for applications from industrial hygiene to leak detection	http://www.raesystems.com/	Photoionization	Minirae3000 Ppbrae-3000 Doserae2 ToxiRAE Pro PID ToxiRAE Pro
RST Rostock system-technik	Early fire detection, applicability in difficult environments, recognition of various detections,	http://www.rst-rostock.de	MOS, QCM	SAMDETECT
Owlstone Nanotech, Inc.,	Pharmaceutical, food freshness and flavor monitoring, detecting various types of chemicals, gases, quality control of spices,	http://www.owlstonenanotech.com/	Ion mass spectrometry	LONESTAR
Illumina	Life sciences, food processing, agriculture, chemical detection	www.Illumina.com	Laser based FO	HiScanSQ Hiseq2500 onose
AltraSens	On-line odor monitoring, Monitoring industrial emissions from a metal processing plant	www.altrasens.de	Quartz microbalance	Odor Vector
Proengin	Detection of chemicals from toxic industrial materials.	www.proengin.com	Flame spectrometry	MAB AP4C, AP2C
Sacmi	Food stuff quality control (wine, olive oil etc.)	http://www.sacmi.eu/	Sensor array	EOS 835 EOS Ambiente

MOS Metal oxide semiconductor, *CP* conducting polymer, *QCM* quartz crystal monitor, *SAW* surface acoustic wave, *IR* infra red, *MOSFET* metal oxide semiconductor field effect transistor, *FO* fiber optic

6.12.2 Smart Sensor

A sensor which provides functions for generating a correct representation of a sensed or controlled quantity is defined as smart sensor. Simplification of the integration of the transducer into applications in a networked environment is the function provided by smart sensor.

As shown in the model of smart sensor (Fig. 6.23), it consists not only transducer (sensing element) but also it contains amplification and signal-conditioning modules, an A/D converter, memory of various types and logic (control) capability are included. In short, a smart sensor is an autonomous digital measuring system equipped with primary information processing, diagnostic and auto-calibration functions and has the ability to communicate with its technical environment. In other words, it is a sensor system integrated with digital blocks that provide digital processing. The block scheme of a simple smart sensor system is depicted in Fig. 6.23.

Value has been added to transducer by smart sensor for data enabling and supporting distributed processing. Interfaces to the transducer, analog processing, nonvolatile memory (stores datasheet information about the sensor), digital processing, and analog or digital communication links contain by a smart sensor.

Table 6.4 shows the summary of advantage and disadvantages of E-nose sensor [18]. Table 6.5 represents the available E-nose instruments in the market.

References

1. P. Nef, How we smell: the molecular and cellular bases of olfaction. Physiology **13**, 1–5 (1998)
2. J.W. Gardner, P.N. Bartlett, *Electronic noses principles and applications* (Oxford University Press, Oxford, 1999)
3. K. Fooladsazl et al., Dopamine determination with a biosensor based on catalase and modified carbon paste electrode with zinc oxide nano-particles. Int. J. Electrochem. Sci. **7**, 9892–9908 (2012)
4. A.K.H. Cheng, D. Sen, H. Yu, Design and testing of aptamer-based electrochemical biosensors for small molecules and proteins. Bioelectrochemistry **77**, 1 (2009)
5. J. Largueze, K.E. Kirat, S. Morandat, Preparation of an electrochemical biosensor based on lipid membranes in nanoporous alumina. Colloids Surf. **79**, 33 (2010)
6. O.A. Sadik, S.K. Mwilu, A. Aluoch, Smart electrochemical biosensors: from advanced materials to ultrasensitive devices. Electrochim. Acta **55**, 4287–4295 (2010)
7. A.P.F. Turner, Biosensors—sense and sensitivity. Science **290**(5495), 1315–1317 (2000)
8. M.K. Beissenhirtz, J. Kafka, D. Schäfer, M. Wolny, F. Lisdat, Electrochemical quartz crystal microbalance studies on cytochrome c/polyelectrolyte multilayer assemblies on gold electrodes. Electroanalysis **17**(21), 1931–1937 (2005)
9. Y. Xiao, F. Patolsky, E. Katz, J.F. Hainfeld, I. Willner, "Plugging into Enzymes": nano wiring of redox enzymes by a gold nano particle. Science **299**, 1877 (2003)
10. M.A. Arnold, M.E. Meyerhoff, Recent advances in the development and analytical application of biosensing probes. Rev. Anal. Chem. **20**, 149 (1988)

11. Y. Mendelson, Optical sensors, in *Encyclopedia of Medical Devices and Instrumentation*, vol. 5, ed. by J.G. Webster (2006), p. 160
12. S. Ismail, Z.A. Ahmad, A. Berenov, Z. Lockman, Effect of applied voltage and fluoride ion content on the formation of zirconia nanotube arrays by anodic oxidation of zirconium. Corros. Sci. **53**, 1156–1164 (2011)
13. L.D. Mello, L.T. Kubota, Review of the use of biosensors as analytical tools in the food and drink industries. Food Chem. **77**, 237–256 (2002)
14. A. Rasooly, K.E. Herold, Biosensors for the analysis of food- and waterborne pathogens and their toxins. J. AOAC Int. **89** (2006)
15. W.J. Kauer, J.S.Dickinson et al., Rapid analyte recognition in a device based on optical sensors and the olfactory system. Anat. Chem. **68**, 2191–2202
16. J. Ito, T. Nakamoto, H. Uematsu, Discrimination of halitosis substance by using QCM sensor array and preconcentrator. Sens. Actuators, B **99**, 431 (2004)
17. K. Persaud, G. Dodd, Analysis of discrimination mechanisms in the mammalian olfactory system using a model nose. Nature **299**(5881), 352–355 (1982)
18. E. Schaller, J.O. Bosset, F. Escher, Food Science Technology. Electronic noses and their application to food. Lebensm.-Wiss. Technol. **31**, 305 (1998)
19. B.C. Munoz, G. Steinthal, S. Sunshine, Conductive polymer-carbon black composites-based sensor arrays for use in an electronic nose. Sens. Rev. **19**(4), 300–305 (1999)
20. I. Lundstrom, S. Shivaraman, C. Svensson, L. Lundkvist, A hydrogen-sensitive MOS field-effect transistor. Appl. Phys. Lett. **26**(2) (1975)
21. S.K. Jha, R.D.S. Yadava, Development of surface acoustic wave electronic nose, Defense Sci. J. **60**(4), 364–376 (2010)
22. I. French, D. George, T. Kretz, F. Templier, H. Lifka, Flexible displays and electronics made in AM-LCD facilities by the EPLaR process. SID-symposium digest of technical papers, vol. 38, pp. 1680–1683 1 May 2007
23. E. Zampetti et al., Flexible sensorial system based on capacitive chemical sensors integrated with readout circuits fully fabricated on ultra thin substrate. Sens. Actuators, B **155**, 768–774 (2011)
24. T. Kinkeldei et al., An electronic nose on flexible substrates integrated into a smart textile. Sens. Actuators, B **174**, 81–86 (2012)
25. Giménez et al., PAMPA III Electronic Nose: Control Electronics Design. J. Argent. Chem. Soc. **93**(1-3), 115–122 (2005)
26. F. Spinelli, M. Noferini, G. Costa, Near infrared spectroscopy (NIRs): perspective of fire blight detection in asymptomatic plant material, in *Proceeding of 10th International Workshop on Fire Blight, Acta Horticulture*, vol. 704 (2006), pp. 87–90
27. S. Sankaran, R. Ehsani, Visible-near infrared spectroscopy based citrus greening detection: Evaluation of spectral feature extraction techniques. Crop Prot. **30**, 1508–1513 (2011)
28. K. Arshak, E. Moore, G.M. Lyons, J. Harris, S. Clifford, A review of gas sensors employed in electronic nose applications. Sens. Rev. **24**(2), 181–198 (2004)
29. H. Troy Nagle, S.S. Schiffman, R. Gutierrez-Osuna, The how and why of electronic noses. IEEE Spectr. **35**(9), 22–34 (1998)
30. D. Zook, U. Bonne, T. Samad, Sensors in control systems. Control Syst. Robotics Autom. **21** (Encyclopedia of life support system (EOLSS))
31. P. Hauptmann, R. Borngraeber, J. Schroeder, J. Auge, Artificial electronic tongue in comparison to the electronic nose—state of the art and trends, in *Proceedings of the 54th Annual IEEE International, Frequency Control Symposium*, USA (2000), pp. 22–29
32. A.D. Wilson, M. Baietto, Applications and advances in electronic-nose technologies. Sensors **9**, 5099–5148 (2009). doi:10.3390/s90705099

Chapter 7
Sensor Circuits

7.1 Introduction

Signal processing is defined as improving the quality of the signal at the output of a measurement system, and very particular is paid to attenuate any noise in the measurement of biosignals that have not been eliminated by the measurement system of the sensor/transducer. Also note that signal processing carries out other functions related to noise, and the exact procedures that are applied depend on the nature of the raw output signal from a measurement transducer. Various types of methodologies including signal filtering, signal amplification, signal attenuation, signal linearization, and bias removal are applied according to the form of correction required in the raw biosignal.

Usually, signal processing is done by analog techniques using various types of electronic circuits. On the other hand, the ready availability of digital computers has meant that signal processing has increasingly been carried out digitally, using software modules to condition the input measurement data.

Digital signal processing is basically more precise than analog techniques, but this advantage is reduced if measurements come from analog sensors and transducers, because an analog-to-digital conversion stage is necessary before the digital processing can be applied. Also, note that analog processing is faster compared to the speed of digital signal processing. So here we discuss analog processing first, then digital processing as well, because analog processing is required along with digital signal processing.

7.2 Analog Conditioning Circuits

The biomedical sensor/e-nose sensor signal is not suitable to be used as an input that is normally sent to a data acquisition system. The original signal might not be suitable because the voltage level is too low, there is some inherent high frequency noise on the signal, or a transformation of the signal must occur before being time

H. K. Patel, *The Electronic Nose: Artificial Olfaction Technology*, Biological and Medical Physics, Biomedical Engineering, DOI: 10.1007/978-81-322-1548-6_7, © Springer India 2014

sampled. Signal conditioning is used to make an optimal match between signal source and the data acquisition system [1].

Proper signal conditioning circuits can improve the quality and performance of a system. Signal conditioning functions are useful for all types of signals, including amplification, filtering, and isolation. They are discussed herewith.

Functions of Signal Conditioning

- Amplification (weak → Strong)
- Attenuation
- Filtering
- Isolation
- Multiplexing
- Digital signal conditioning

Amplification

E-nose sensor signals are very small in magnitude; signal conditioning can improve the accuracy of sensor data. Amplifiers boost the level of the input signal to match the range of the analog-to-digital converter (ADC), thus increasing the resolution and sensitivity of the measurement. To increase the resolution of the input signal and signal-to-noise ratio (SNR) are the main requirements of ADC. Amplification of signal is shown in Fig. 7.1.

Attenuation

Attenuation is just working on the opposite of amplification. It is required when the voltages to be digitized are beyond the input range of the digitizer. These forms of signal conditioning reduce the amplitude of the input signal so that the conditioned signal is within range of the ADC. Attenuation is necessary for measuring high voltages. Signal attenuation is shown in Fig. 7.2.

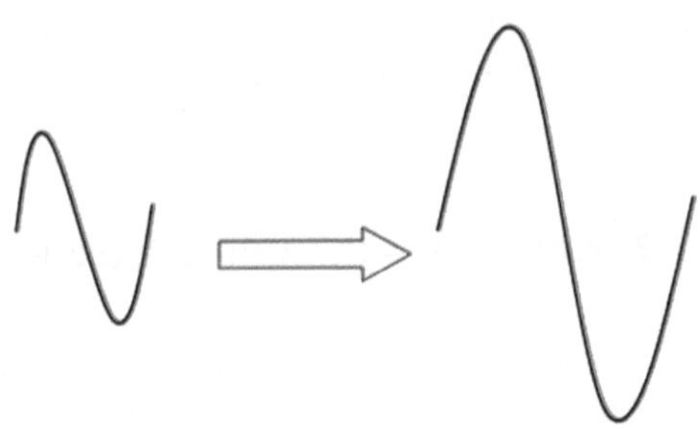

Fig. 7.1 Amplification of signal

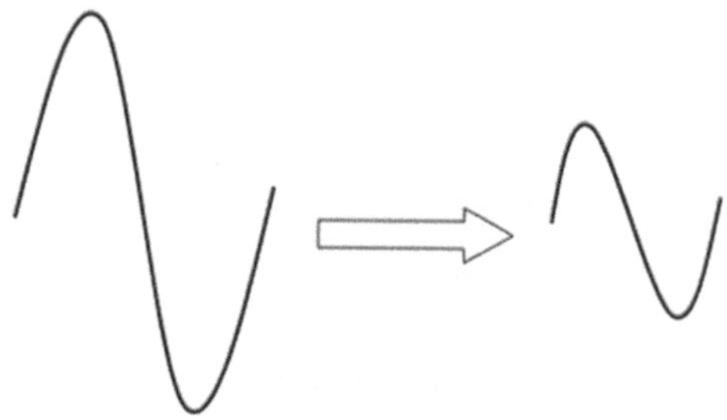

Fig. 7.2 Attenuation of signal

Filtering

To reject unwanted noise within a certain frequency range, signal conditioners can include filters. A further common use of filters is to prevent signal aliasing—a phenomenon that arises when a signal is undersampled (sampled too slowly).

Isolation

Improper grounding of the system is one of the most common problems which include noise and cause damage to measurement devices. Signal conditioners with isolation can help to prevent most of these problems. Isolation functions include: breaking of ground loops, rejection of high common-mode voltages, and protection of expensive instruments. Optical, Magnetic, and capacitive isolators are used for circuit isolation and these devices pass the signal from its source to the measurement device without a physical connection. Signal is modulated by using magnetic and capacitive type isolators from a voltage to a frequency. Before converting back to a voltage value the frequency can be transmitted across a transformer or capacitor without direct physical connection.

Multiplexing

A basic method used for measuring various signals with a single ADC is multiplexing. The ADC functions alike: it samples one channel, toggles to next channel and samples it, and so on. As the same ADC is sampling many channels rather than only one, the efficiency rate of each individual channel is proportional to the number of channels sampled. The digitizer is the most expensive part of a data acquisition system. By use of multiplexing, it can sequentially route a number of signals into a single digitizer, thus achieving a cost-effective way to greatly expand the signal count of system. Signal conditioners set with signal multiplexers can cost-effectively expand the input/output (I/O) capabilities.

Digital Signal Conditioning

Digital signals can require signal conditioning peripherals.

List of analog circuits used as components of signal analysis/conditioning of biosensors.

1. Conditioning bridge circuit
2. Amplifier for signal conditioning circuit
   - Inverting amplifier
   - Summing amplifier(extension of inverting)
   - Non-inverting amplifier
   - Differentiator
   - Comparator
   - Rectifier
   - Integrator
3. Filter circuit
   - Active
   - Low pass
   - High pass
   - Band pass
4. ADC for signal conditioning circuit

1. **Conditioning Bridge Circuit**

Sensor outputs are often primarily in some non-voltage form; conversion to a measurement signal that is in a more convenient form can be achieved by various types of variable conversion elements in the measurement system. Bridge circuits are a particularly important type of variable conversion elements. Bridge circuits are among the basic fundamental and powerful electrical tools found in measurement, switching, oscillator, and transducer circuits. Bridge circuit is an accurate method of measuring resistance, inductance, and capacitance values, and enables the detection of very small changes in these quantities of about a nominal value.

$$V_0 = V_B\left(\frac{R1}{R1+R4} - \frac{R2}{R2+R3}\right) \text{at balance,} \quad V_0 = 0\,\text{if}\,\frac{R1}{R4} = \frac{R2}{R3}$$

Basically null and deflection types of bridge exist; null types are employed for calibration purposes (balanced mode) and deflection types (unbalance mode) are used within closed-loop automatic control schemes. Bridge is an attractive alternative to the potential divider.

A null-type bridge with d.c. excitation, commonly known as a Wheatstone bridge, has the form shown in Fig. 7.3. The bridge contains four resistors connected to a quadrilateral, a source of excitation voltage defined as V_B (or, alternately, a current) connected across one of the diagonals of the bridge, and a voltage detector connected across the other diagonal of bridge. The detector measures the difference between the outputs of the two voltage dividers connected

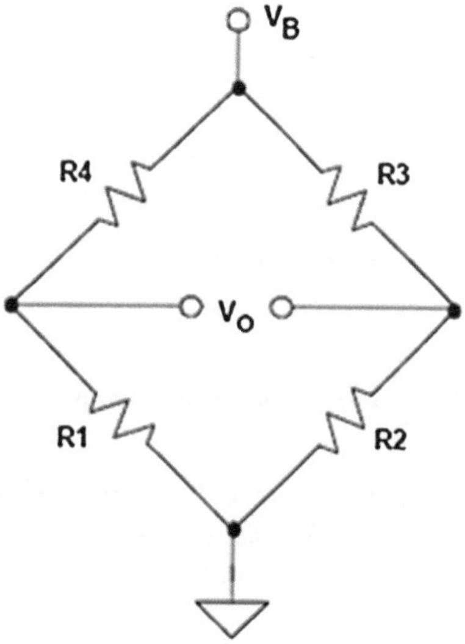

Fig. 7.3 Wheatstone bridge

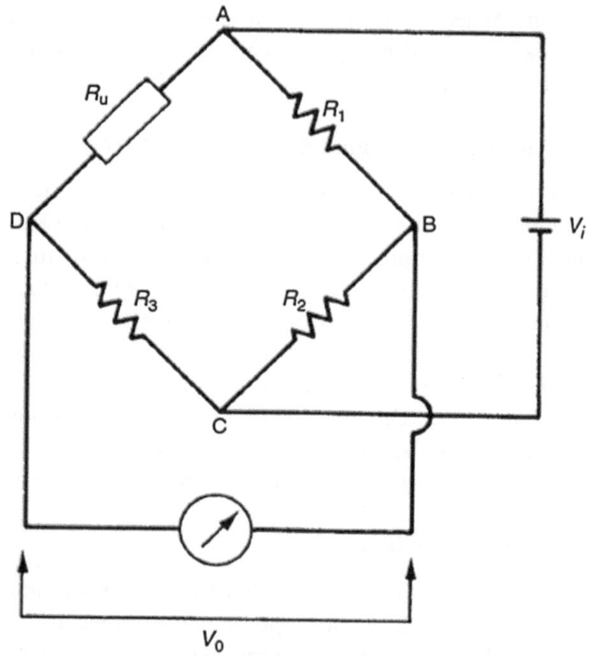

Fig. 7.4 Deflection type DC bridge

across the excitation voltage, V_B. The general structure of the bridge circuit output V_O is illustrated in Fig. 7.3.

A deflection-type bridge with d.c. excitation is shown in Fig. 7.4. This differs from the Wheatstone bridge mainly in that the variable resistance R_v is replaced by

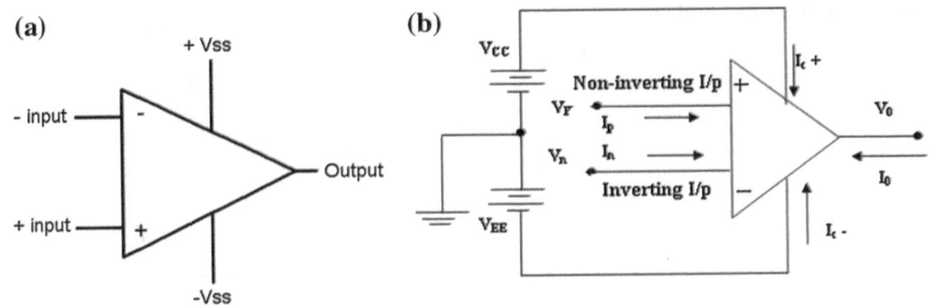

Fig. 7.5 (**a**) Basic op-amp (**b**) op-amp circuit

a fixed resistance R_1 of the same value as the nominal value of the unknown resistance R_u. As the resistance R_u changes, the output voltage V_O varies and this relationship between V_O and R_u must be calculated.

$$V_O = V_i\left(\frac{R_u}{R_u + R_3} - \frac{R_1}{R_1 + R_2}\right)$$

The Wheatstone bridge configuration is used in many sensor circuit analysis technologies and also specifically bridge gas configuration applied in gas sensor analysis technology.

2. **Amplifier for Signal Conditioning Circuit**

An amplifier is simply a device that increases the magnitude of voltage, current, or power. It is the basic building block of Analog Electronic Circuits which is used in signal conditioning, filtering or to perform mathematical operations. Op-amp is normally used in closed-loop mode for amplifying and signal conditioning. High input impedance and low output impedance and high gain differential amplifier is a basic characteristic of operational amplifier.

The operational amplifier (op-amp) is a voltage controlled voltage source with very high gain. Op-amp has two input terminals, one output terminal, two power supply terminals, and a ground connection. The symbol of the op-amp with the associated terminals and output terminal (ports) is shown in Fig. 7.5a, b.

Inverting Amplifier

The basic working principle of the inverting amplifier as shown in Fig. 7.6 is that it simply accepts an input signal referenced to common and amplifies it and inverts the polarity at the output terminals. The input impedance of the inverting amplifier is finite. In short the amplifier gives a negative gain to the input voltage, so output is out of phase with input voltage.

$$\boldsymbol{\beta} = -\frac{\boldsymbol{R}_1}{\boldsymbol{R}_F}\,\mathrm{So}, \boldsymbol{A} = \frac{\boldsymbol{A}_0}{1 - \boldsymbol{A}_0\frac{\boldsymbol{R}_1}{\boldsymbol{R}_F}} = \frac{\boldsymbol{R}_F}{\frac{1}{A_0}\boldsymbol{R}_F - \boldsymbol{R}_1} \cong -\frac{\boldsymbol{R}_F}{\boldsymbol{R}_1} = \frac{1}{\boldsymbol{\beta}}$$

Fig. 7.6 Inverting amplifier

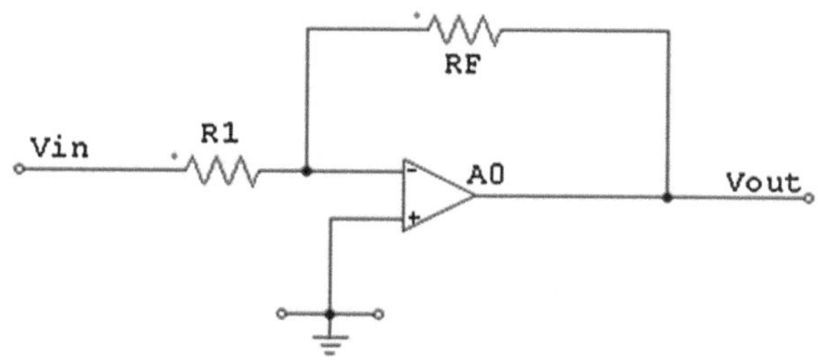

Non-Inverting Amplifier

The signal input is applied to the non-inverting (+) input. The inverting input (−) is grounded. The resistor R_2 is the feedback resistor. It is connected from the output to the positive (inverting) input. This circuit as shown in Fig. 7.7 can be used to buffer the output of a sensor from the load effect of the next stage and to introduce a gain of greater than 1 to the signal. The input impedance of the non-inverting amplifier is infinite.

$$\frac{e_0}{e_i} = \frac{i(R_2 + R_1)}{iR_1} = \frac{R_2 + R_{1i}}{R_1}$$

Summing Amplifier (Extension of Inverting)

A common modification of the inverting amplifier is an amplifier as shown in Fig. 7.8a, b that sums or adds two or more applied voltages.

$$V_{\text{out}} = \left(1 + \frac{R_F}{R_a}\right)\left(\frac{R_A}{R_a}V_a + \frac{R_A}{R_b}V_b + \frac{R_A}{R_c}V_c\right) \text{where } R_A = \frac{1}{\frac{1}{R_a} + \frac{1}{R_b} + \frac{1}{R_c}}$$

$$\text{if } R_a = R_b = R_c \text{ then } V_{\text{out}} = \left(1 + \frac{R_F}{R_B}\right)\left(\frac{V_a + V_b + V_c}{3}\right)$$

Differential Amplifier

The inverting and non-inverting properties of an op-amp can be combined together in a differential amplifier. This circuit as shown in Fig. 7.9 is used to amplify the difference between two input signals. The two inputs are connected to

Fig. 7.7 Non-inverting amplifier

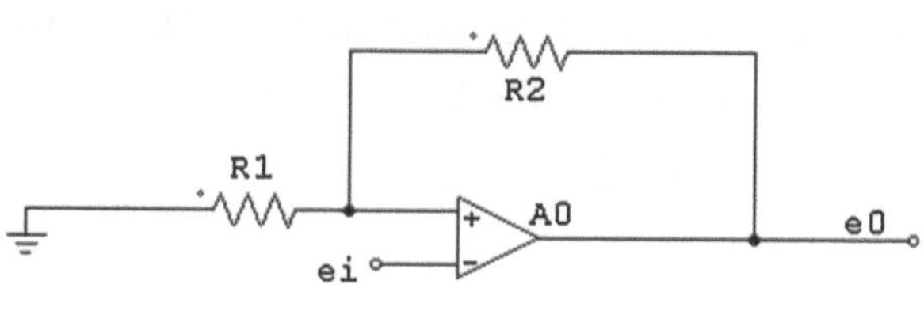

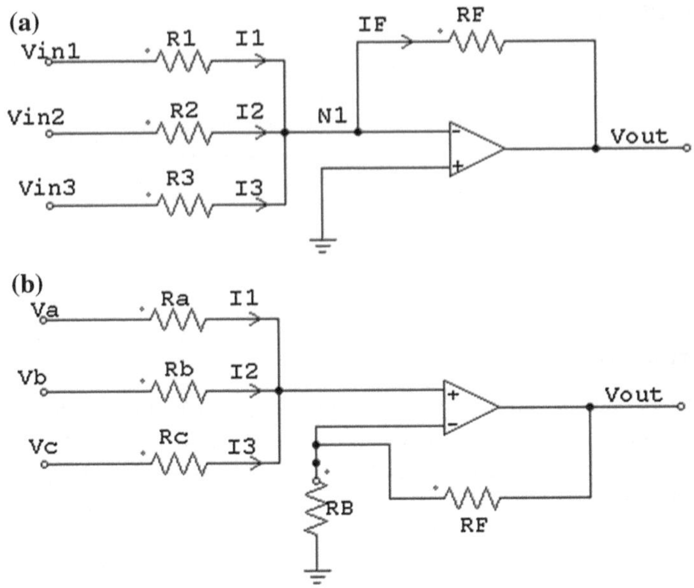

Fig. 7.8 **a** Summing amplifier. **b** Summing amplifier

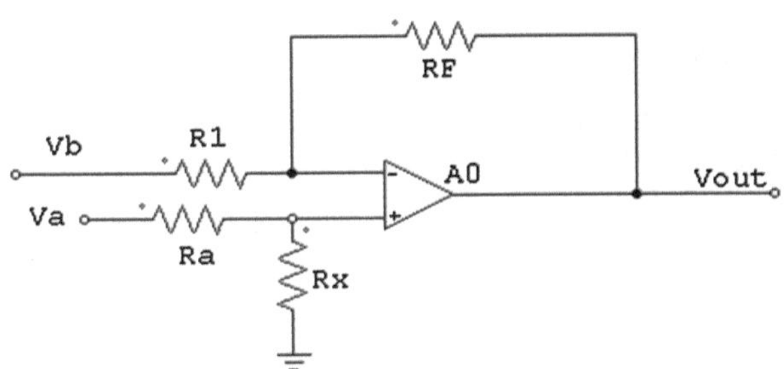

Fig. 7.9 Differential amplifier

two independent sources. A differential amplifier is commonly used as a noise rejection circuit (i.e., sensitive signals need to be amplified, when signal-carrying cables may pass through noisy environments, where signal-carrying cables may be extremely long, etc.). This circuit is useful in high-pass filter circuits. Each signal connection may be affected by noise or "pick-up," but if they closely follow the same path (being twisted together, for example), they will pick up the same amount of noise. Integrity of sensitive or low-level signals may be compromised by the use of general "ground" as one "side" of the signal connection.

$V_{out} = -\frac{R_F}{R_1} V_b + \left(1 + \frac{R_F}{R_1}\right)\left(\frac{R_x}{R_x + R_a}\right) V_a,$ If $R_a = R_1$ and R_F then $\boldsymbol{V_{out} = (V_a - V_b)\frac{R_F}{R_1}}$

Voltage Comparator

A comparator compares the voltages at the (+) and (–) inputs. If the (+) input is at a higher voltage than the (–) input the comparator output will be high. If the (–) input is at a higher voltage than the (+) input the comparator output will be low (Fig. 7.10).

The circuit diagram of an integrator using op-amp is shown in Fig. 7.11. The output voltage is the time integral of the input voltage as shown in the waveform figure.

In multiple sensor array analysis buffer op-amp provides isolation between two parts of the circuit or works as buffer in between two circuit analysis with impedance matching problem. These buffer circuits as shown in Fig. 7.12 typically have a gain of 1 but have very large input impedance and very small output impedance. Because their gains are 1, they are also called voltage followers.

Rectifier circuits are mostly used in the design of power supply circuits. Further applications are also there, i.e., in instrumentation applications, the signal to be rectified can be of incredibly small amplitude, making it impractical to use the conventional rectifier circuits. Also it is used for very precise transfer characteristics. Hence, it is also called as precision rectifier.

A half-wave precision rectifier is constructed with the use of an op-amp, and includes the diode in the feedback loop (as shown in Fig. 7.13). This efficiently terminates the forward voltage drop of the diode, so very low-level signals (well below the diode's forward voltage) can still be rectified with minimal error.

Full-wave rectifiers are more difficult, compared to the half-wave circuits. Full-wave rectifiers output one polarity of the input signal and reverse the other. A circuit for a full-wave rectifier is shown in Fig. 7.14.

3. **Filter**

Filtering is the process of removing certain portions of the input signal in order to create a new signal. Filters can be broadly grouped into four classes according to their frequency response.

- Low-pass,
- High-pass,
- band-pass, and
- band reject (notch).

These analog filters can be implemented using passive or active circuits. Passive filters consist of networks of resistors, capacitors, and inductors, whereas active filters contain active components (e.g., op-amps, transistors), resistors, and

Fig. 7.10 Voltage comparator

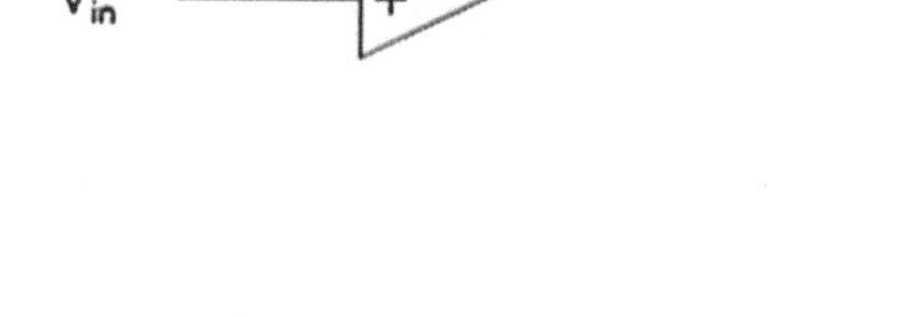

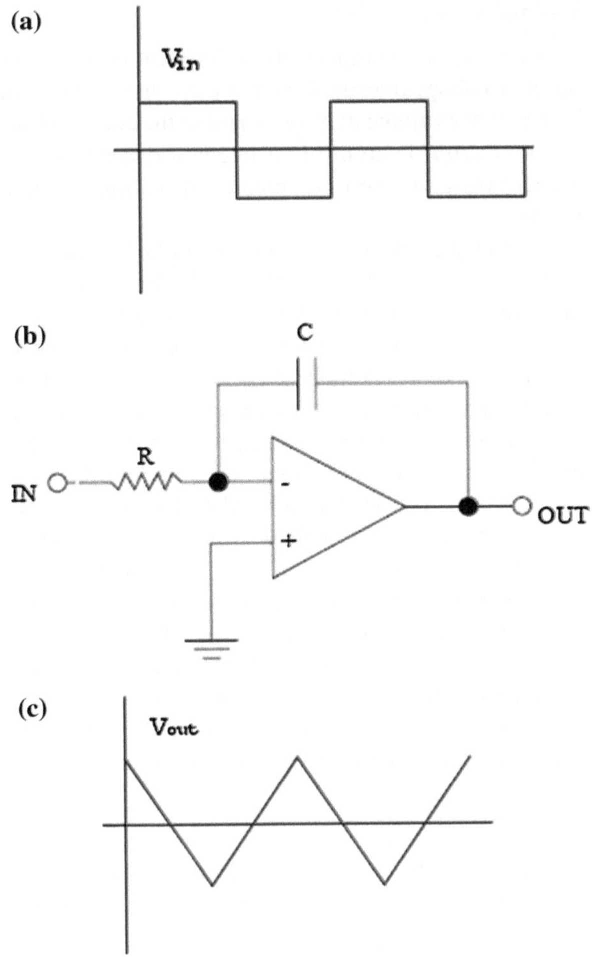

Fig. 7.11 Integrator

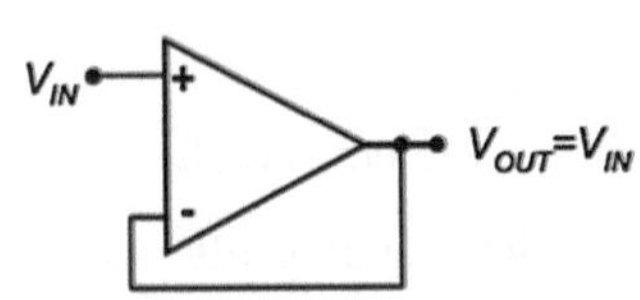

Fig. 7.12 Voltage follower

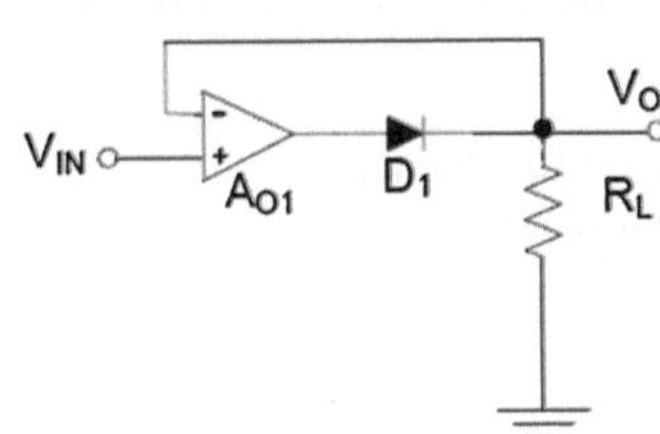

Fig. 7.13 Op-amp used as half wave rectifier

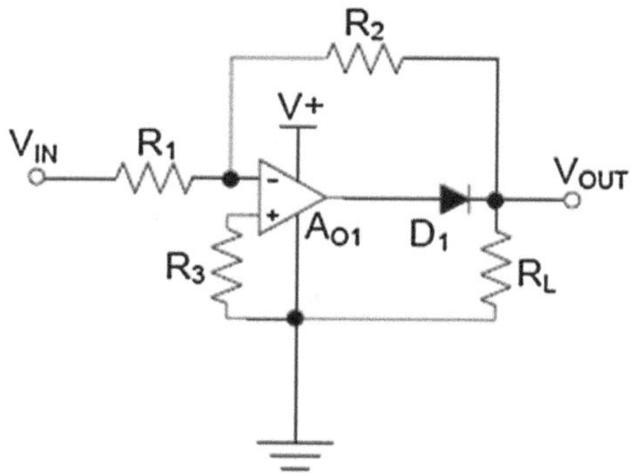

Fig. 7.14 Full wave rectifier circuit using op-amp

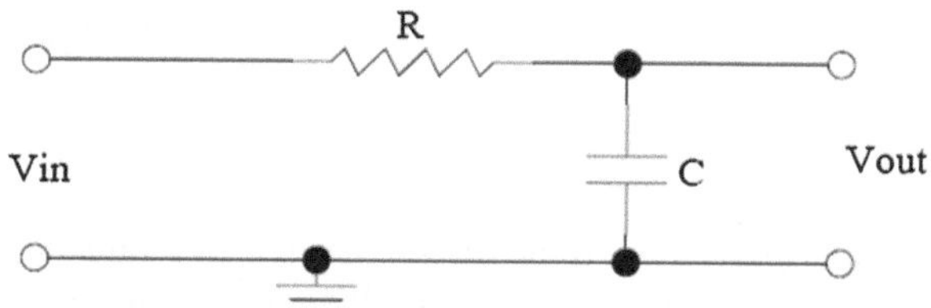

Fig. 7.15 Low pass R–C filter

capacitors. Active filters are suitable for low frequency, small signals but require power supply to operate. Passive filter has the added advantage of low noise. An active filter is also based on op-amp circuits.

- The filter which removes all frequencies above the cut off frequency is defined as low-pass filter. (Typically used for noise removal and data smoothing.) A low-pass R–C filter circuit is shown in Fig. 7.15.

The high-pass filter removes all frequencies below the cut-off frequency. (Used for DC or low-frequency drift). The high-pass filter is designed to contain a lower corner frequency close to zero and the cut-off frequency at near to several higher values. The number of capacitor/resistors joining up decides the number of poles and the degree of cut-off sharpness. It also operates exactly reverse to low-pass filter. Also note that a combination of high-pass and low-pass filters may be used to create a notch filter to attenuate a narrow band of frequencies. Figure 7.16 demonstrates the circuit of high-pass filter.

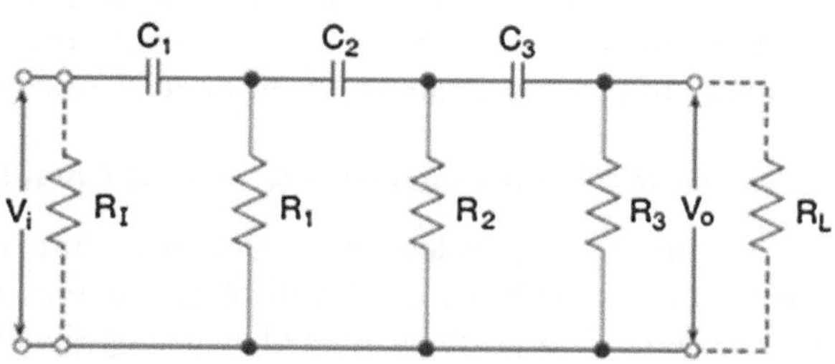

Fig. 7.16 High pass filter

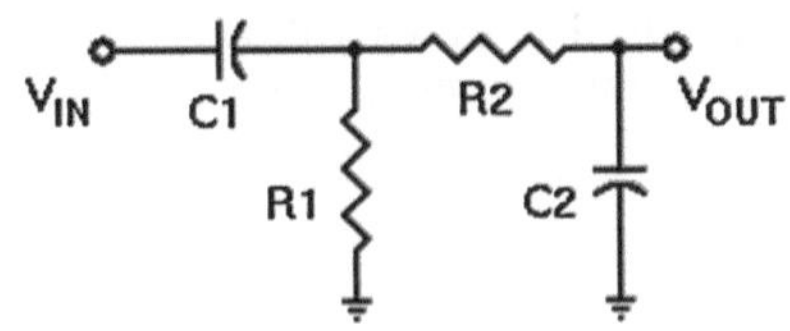

Fig. 7.17 Band pass filter

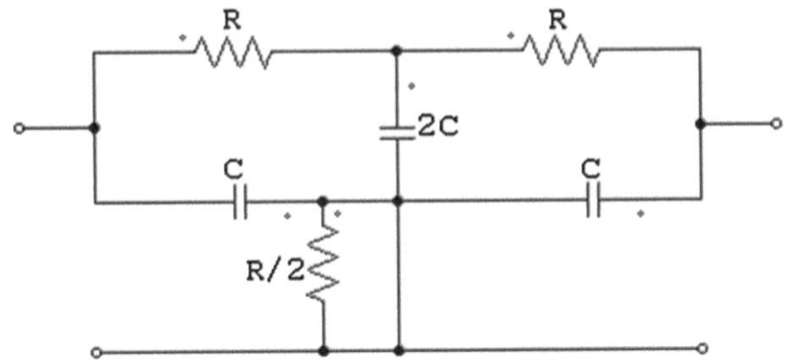

Fig. 7.18 Band reject filter (Notch filter)

- Band-pass filter removes all frequencies outside (f_1–f_2). (Used often in EEG measurements) A series combination of the low-pass filter and the high-pass filter results in a band-pass filter, which amplifies frequencies over a desired range and attenuates higher and lower frequencies. (A band-pass filter attenuates both low and high frequencies). A schematic of band-reject filter is shown in Fig. 7.17.
- Band reject or notch filter removes all frequencies between f_1 and f_2. (60 Hz noise removal). A schematic of band reject pass filter is shown in Fig. 7.18.

This notch filter can be helpful in rejecting unwanted signals that are on a particular frequency. The response provided by the filter includes a low level of attenuation away from the notch frequency. As a signal shifts nearer to the notch frequency, the level of attenuation rises, giving the typical notch filter response. Response of various types of filters is illustrated in Fig. 7.19.

When during the signal analysis, the physical quantity being measured is either constant or only changing slowly with time, at that time the system noise frequency is high. So there is a requirement of low-pass filter. In a few cases, the measured signal itself has a high frequency, for instance when mechanical vibrations are being monitored, and the signal processing required is the application of a high-pass filter to attenuate low-frequency noise components. Band-stop filters can be used where a measurement signal is corrupted by noise at a particular frequency. Such noise is frequently due to mechanical vibrations or proximity of the measurement circuit to other electrical apparatuses.

4. **Analog-to-Digital Converters for Signal Conditioning Circuit**

The conversion of analog signals into digital form is essential in any system that will use digital methods, i.e., for display, counting, or any other logical actions. (Conversion means the processing of an analog signal into a set of digital signals).

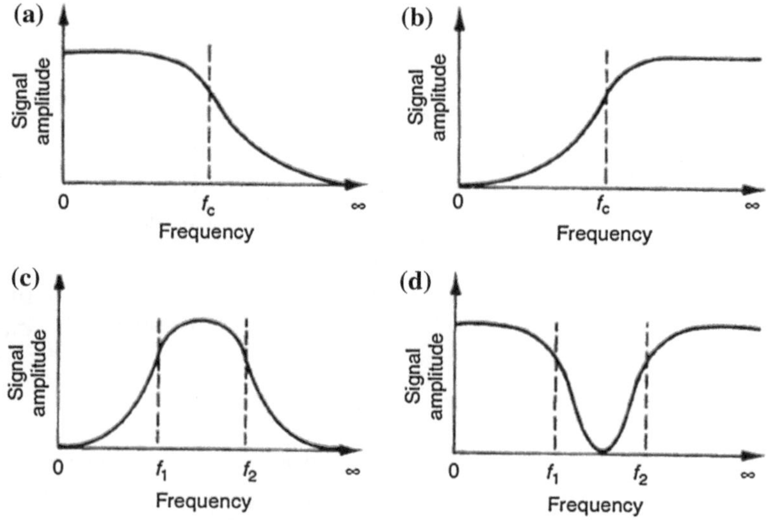

Fig. 7.19 Response of filter **a–d** Frequency

Digital to Analog Converter (DAC) and Analog to Digital Converter (ADC) work as the interface between digital/analog devices. They are also defined as mixed-signal devices as they perform both analog and digital functions.

There are many methods available to convert the analog signal in to digital signal. They are counter, single slope ADC, dual slope ADC, Flash, delta and sigma, and successive approximation ADC. In particular, ADC is an electronic integrated circuit which transforms a signal from analog (continuous) to digital (discrete) form. As digital forms of signals are less susceptible to the deleterious effects of additive noise, ADC is required.

ADC provides a link between the analog transducers, the digital signal processing, and data handling. Almost everywhere ADC is used where an analog signal has to be processed, stored, or transported in digital form. Some examples of ADC applications are digital volt meters, cell phone, thermocouples, and digital oscilloscope. Comparison of different types of ADCs is shown in Table 7.1 and the basic process of ADC is illustrated in Fig. 7.20.

Quantizing: In binary partitioning the reference signal ranges to a number of discrete quanta, then matching the input signal to the correct quantum.

Table 7.1 ADC Types: comparison

Type	Speed (Relative)	Cost (Relative)
Dual slope	Slow	Medium
Flash	Very fast	High
Successive approximately	Medium-fast	Low
Sigma-delta	Slow	low

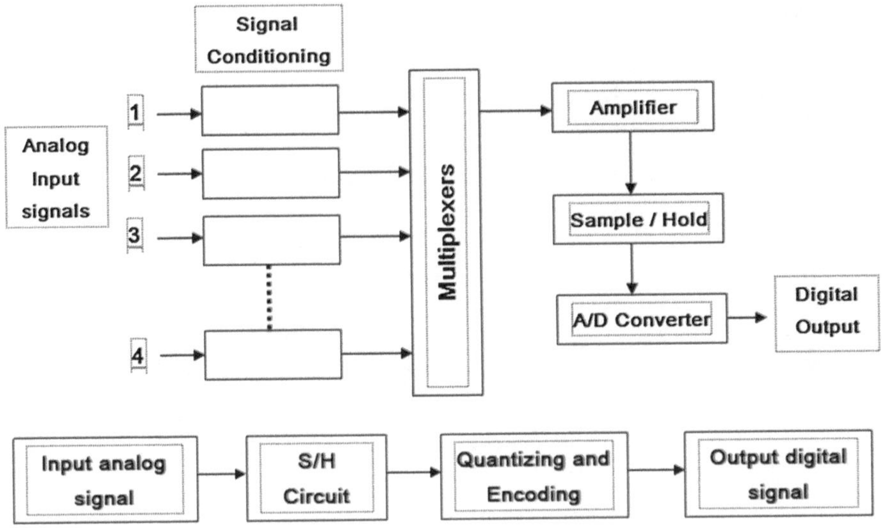

Fig. 7.20 ADC process

Multiplexer (MUX)—a switching device that sequentially connects multiple inputs or outputs in order to process several signal channels with a single A/D or D/A converter.

Encoding: Assigning a unique digital code to each quantum, then allocating the digital code to the input signal. ADC is also called encoding device.

Sample-and-Hold (S/H) Circuit—it acquires and stores an analog voltage on a capacitor for subsequent processing.

Note that the ADC also works as a divider because ADC has an analog reference voltage or current next to which the analog input is compared.

A digital to analog converter (DAC) has a digital input that specifies an output whose value changes in step. These step changes are in volts or amperes. An input that can vary from a minimum to a maximum value of volts or amperes is the input of the analog to digital converter.

The output is a digital number that represents the input value. Digital encoder (8-line to 3-line) is shown in Fig. 7.21.

A DAC takes an n- bit digital input and outputs a corresponding analog voltage. DAC systems normally consist of three components: (a) A reference voltage, (b) The DAC itself, and (c) An op-amp for output buffering. DAC is also called decoding device.

Data Analysis Techniques

The sensor signals obtained are saved into the computer and then the data analysis process is started. Signal preprocessing is the first computational stage, which fulfills various purposes which include compensating for sensor drift, extracting

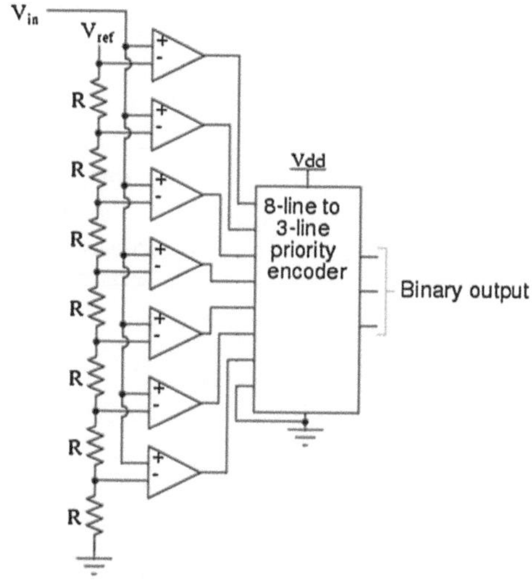

Fig. 7.21 Digital encoder

descriptive parameters from the sensory array response, and preparing the feature vector for further processing. Improvements in the performance of the available data analysis techniques are an important topic of research on electronic nose development. Figure 7.22 represents the schematic process presentation in data analysis techniques.

Processing may do the following things.

- Change the level or value of the signal (e.g., voltage level).
- Change the signal from one form to another (e.g., current to pneumatic).
- Change the operating characteristic with respect to time.
- Convert analog and digital signals from one to the other.

Preprocessing Techniques

The raw data were preprocessed to improve the quality of the input data. Data preprocessing techniques are designed to provide modification of existing raw data in order to achieve more quality supply in input variable.

Prior to any data analysis, it is usual to carry out preprocessing of the data. The main aims of this stage are:

1. To reduce the amount of data those are not related to the study.
2. To increase sufficient information within the data to get the desired purpose.
3. To remove the information in, or transform the data to, a form suitable for further analysis.

Some examples of signal processing algorithms used are shown in Table 7.2 [2]. (i = sensor, j = odor, a = odor a, b = reference odor b, σ = population standard deviation, X=average value, N = the number of feature vectors in the feature-set with i component to each vector).

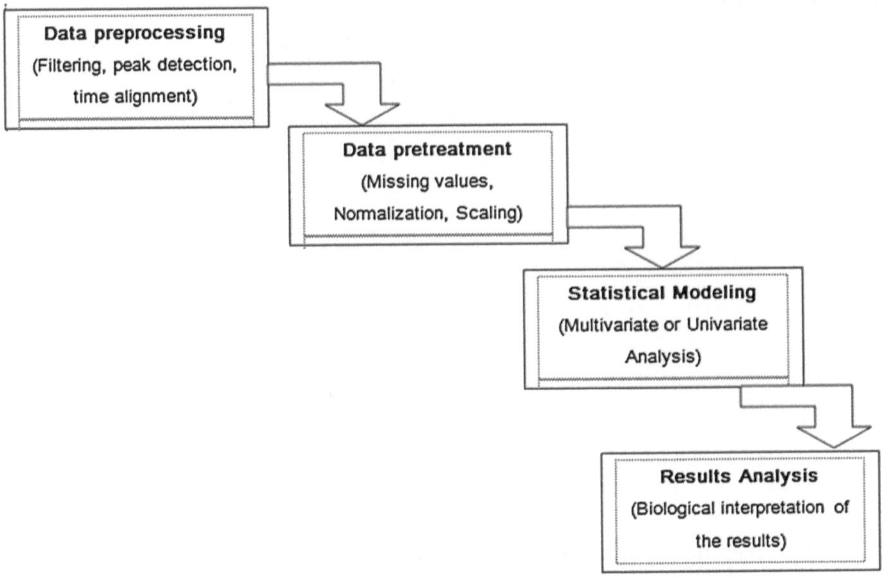

Fig. 7.22 Schematic of process in data analysis techniques

The selection of signal preprocessing is important and also it has a significant impact on the performance of subsequent module in the pattern analysis system. Although signal preprocessing is somewhat dependent on the underlying sensor technology, the most commonly used three general stages can be identified: baseline manipulation, compression, and normalization.

Table 7.2 Signal processingalgorithm [2]

Signal processing algorithms	Formula
Difference signal	$X_{ij} = X'_{ij} - X'_{b,ij}$
Relative signal	$X_{ij} = X'_{a,ij} / X'_{b,ij}$
Averaging	$\underset{ij}{\bar{X}} = \sum_{j=1}^{N} X'_{ij} / N$
Fractional difference	$X_{ij} = (X'_{a,ij} - X'_{b,ij}) / X'_{b,ij}$
Linearization	$X_{ij} = \log\left(X'^{\max}_{ij} - X'^{\min}_{ij}\right)$
	$X_{ij} = \sqrt{X'^{\max}_{ij} - X'^{\min}_{ij}}$
Normalization	$k_{ij} = X_{ij} / \left\lvert X'_{a,ij} - X'_{b,ij} \right\rvert,$
	$k_{ij} = X_{ij} / \sum_{i=1}^{m} X^2_{ij}$
Autoscaling	$k_{ij} = \frac{\left(x_{ij} - \underset{ij}{\bar{X}}\right)}{\sigma_i}$

Baseline Manipulation

This set of operations refers to transformations based on the initial value of the transient (the baseline) and is an attempt to reduce the effects of sensor drift [3]. The primary stage of preprocessing consists of manipulating the sensor response with respect to its baseline (e.g., response to a reference analyte) for purposes of drift compensation, contrast enhancement, and scaling.

The choice of baseline manipulation technique and response parameter (e.g., resistance, conductance, frequency) is very much reliant on the sensor technology and the particular application where it is implemented [4].

Normalization

Normalization is a method to eliminate quantitative information from a data set. The vector length (magnitude) is represented by the concentration of a variable [5]. Each sensor response is normalized for qualitative assessment of the original data set according to:

$$X'_{ij} = X_{ij} \Big/ \sqrt{\sum_{k=1}^{m} X_{ik}^2}$$

where x'_{ij} = the normalized sensor response, x_{ij}= the raw sensor response, and x_{ik}= the individual sensor responses. Dividing each sensor response by the sum of all the sensor responses to the same sample such that the concentration information disappears is defined as normalization.

Normalization is generally useful to neutralize possible variations in the sample concentration. These variations are generally due to sample temperature variations, and to instabilities of the sampling system that may lead to variations of the dilution factor of the sample with the carrier gas. Normalization has partial efficiency because the mentioned theory is only used for simple gases and it is not used when mixtures of compounds are measured. Also, in complex mixtures, temperature fluctuations do not result in a general concentration shift, but since individual compounds have different boiling temperatures, each component of a mixture changes differently so that both quantitative (concentration shift) and qualitative (pattern distortion) variations occur.

Scaling

Generally, raw data cannot be used directly; instead two main scaling procedures are widely used: zero-centered and auto scaling. Zero-centered data means that each sensor is shifted across the zero value so that the mean of the responses is zero. Compared to sensors showing large changes, often, sensors with small signal changes are equally important. For solution of the said problem the input variable can be scaled either to similar ranges defined as range scaling, or to similar variances defined as auto scaling. To eliminate the influence of absolute values by scaling them to values between zero and one is a procedure order in range scaling [6].

Auto scaling means to scale each sensor to zero-mean and unitary variance. This operation equalizes the dynamics of the sensor responses avoiding that a sensor with a larger response range may hide the contribution of other sensors dynamically limited. *Auto scaling* is a technique to scale the variables to related standard deviations and so reduce the effect of signal changes (variance). Further, auto scaling makes the sensor responses dimensionless; this feature becomes necessary when sensors whose signals are expressed in different units are joined in the same array. This is the case of hybrid arrays (different sensor technologies in the same array) and when electronic noses are fused with other instruments, e.g., the fusion of electronic noses and electronic tongues.

There are a number of pattern-recognition methods used to analyze the response produced by sensor array [7]. Pattern recognition for odor analysis demands statistical/mathematical tools and related software as the sensors inherently exhibit highly nonlinear characteristics upon introduction to gases with dispersion occurring at high concentration.

Pattern-recognition techniques are used for data processing of output responses generated by each sensor of e-nose system. The basic advantages of this method include the reduction in difficulty of the sensor coating selection, and the capability to distinguish complex mixtures without the need to identify and quantify individual components. In this section, we will briefly review the concepts behind this method.

Pattern recognition is a decision vector used to classify class based on a series of patterns on that class. Usually a matrix is formed from the patterns for a number of classes and then a decision vector which divides the pattern into an assigned binary classification is calculated based on standard experiments. This is then used to categorize unknown patterns. The success of Pattern-recognition techniques can be improved or simplified by suitable prior handling of the data such that feature selection and feature extraction are important approaches [8]. Pattern-recognition methods are mainly divided into supervised and non-supervised methods, although a combination of both can be used. The major unsupervised technique is principal component analysis (PCA) while artificial neural network (ANN) is the best-known supervised technique [6].

The signal recorded and the pattern for each sample can make an exploratory analysis with all the information that is required. The most important multivariable tool for exploratory analysis is Principal Component Analysis (PCA). Using the PCA the measured data will be transformed into 2D or 3D coordinates. PCA is a linear supervised pattern-recognition technique that has often been used to explore gas sensor array data in conjunction with cluster analysis.

A set of principal components are transformed from a set of correlated variables such that the first few components define most of the variation in the data set as the methodology followed in PCA. It is a specific kind of orthogonal projection and its coordinate system is usually called 'feature space'. A method which is defined as an unsupervised data reduction method is PCA.

The principle function of the PCA method is to illustrate variations of a multivariate data set in terms of a set of uncorrelated variables, each of which is a particular linear combination of the original variables. In short, PCA is a linear feature extraction technique which shrinks the dimensionality of data with the least loss of information. This technique is achieved by projecting the data onto fewer dimensions which are chosen to utilize the relationships between the variables, so that the maximum amount of information is retained in the smallest number of dimensions. This technique allows the similarities and differences between samples to be better reviewed.

Exploratory (Multivariate) Techniques

The most widespread multivariate tool for exploratory analysis is still Principal Component Analysis (PCA). There are sets of data given related to a number of measurements, multivariate techniques that aim at studying the intrinsic characteristics of the data to discover data internal properties. Multivariate analysis supports the attitude of an electronic nose to be utilized for a specified application, leaving to the supervised classification the task of building a model to be used to predict the class membership of unknown samples. Two main groups of multivariate analysis may be identified: dimensionality reduction techniques and clustering techniques.

Considering and understanding multivariate data is the key to success. The available list of multivariate methods is long (as we listed same in Chap. 5), but only a limited number of methods are normally used in gas sensing. A group of algorithms aimed at providing a representation of the data in a space of dimensions lower than the original sensor space is defined as dimensionality reduction technique.

All techniques are based on specific propositions about the nature of the data and the sensor space. Each technique is responsible to maintain some particular and defined characteristic of the data. The simplest, calculus, and interpretation of results are based on the strongest assumption about the statistical distribution of the data. A neural network is necessary for data representation where assumption of data distribution is removed.

Clustering techniques are generally based on the theory of connection expressed through the definition of a metric (distances calculus rule) in the sensor space. The most insignificant and common choice is to express the similarity as a Euclidean distance [9].

Principal Component Analysis

From neuroscience to computer graphics a simple, non-parametric method of all types of data analysis is one and only Principal Component Analysis (PCA). PCA is required for information extracting from confusing data sets. PCA is a method to recognize patterns in data, and state the data in a way to highlight their similarities and dissimilarities [10]. In simple ways, PCA is a numerical method for analyzing the basis of variation present in a multidimensional data set [11]. From applied linear algebra most valuable results have been described by PCA. For reducing complex data sets to a lower dimension to reveal many hidden, simplified structure that often are most useful, PCA is implemented with minimal further efforts. An unsupervised data reduction method which is identified as one of the most successful techniques is illustrated as PCA.

To describe variation of a multivariate data set in terms of a set of uncorrelated variables is the basic principle of the PCA data analysis methods. In PCA each of the uncorrelated data variables is a particular linear combination of the original variables. Simply said in another way, the original data matrix is projected from a high-dimensional space into a less-dimensional space, specifically a plane or a three-dimensional space. In the process stage the original data set is reduced in dimension or compressed which results in little loss of information.

A basis of a space which is represented by training vectors is calculated by the PCA and this basis vectors are actually eigenvectors. Eigenvectors are computed by PCA in the direction of the largest variance of the training vectors. These basis vectors, actually eigenvectors, computed by PCA are in the direction of the largest variance of the training vectors. Consideration of quantitative measurement of how much a component represents the data represents eigenvalues. As the eigenvalues of a component are higher, it represents more of the data. For simplest and robust ways of dimensionality reduction PCA is the best solution [9].

PCA is generally related with identifying correlations in the data (Correlation measures the simultaneous change in the values of two or more variables). There are various models available for describing the behavioral nature of a simultaneous change in values, such as linear, exponential, periodic, and more. In PCA the correlation used is linear.

When obtained from procedures on a number of pragmatic variables and development of smaller number of artificial variables identified as principal components a suitable analysis technique is PCA. The principal components may then be used as analyst or standard variables in subsequent analyses. Principal component analysis is a variable reduction method. It is useful when collecting data on a number of variables (possibly a large number of variables), and considering that there are few redundancies in those variables.

The main functions of PCA are prediction, redundancy removal, feature extraction, data compression, etc. Applications such as signal processing, image processing, system and control theory, communication, etc., have linear models and appropriately implement a traditional technique PCA.

There is an option of a reliable representation of an e-nose data set in a subspace of reduced dimensionality produced with the fact that the chemical sensors always exhibit a certain degree of correlation among them. The principal component analysis consists in finding an orthogonal basis where the correlation among sensors disappears.

Artificial Neural Network

The main characteristic of an Artificial Neural Network (ANN) model is whether it needs direction in learning or not. Based on the way they learn, all artificial neural networks can be divided into two learning categories: supervised and unsupervised. In supervised learning, a preferred output result for each input vector is necessary when the network is trained. An ANN of the supervised learning kind, such as the multilayer perception, uses the objective result to conduct the formation of the neural parameters. Hence to learn the behavior of the process under study, it is possible to build the neural network.

In unsupervised learning, the working out of the network is totally data driven and no target results for the input data vectors are provided. An ANN of the unsupervised learning type, such as the self-organizing map, can be utilized for clustering the input data and find features inherent to the problem. The ANNs are computer programs based on a simplified model of the brain; they reproduce its logical operation using a collection of neuron-like entities forming networks to perform processing. ANN programs are multipurpose and with suitable training, a single program could solve a number of problems [9].

A neural network required two physical components, (1) the processing elements (neuron) (2) the connection between the elements (links).Every link has a load parameter associated with it.

Each neuron processes the information and produces the output from the stimulus which they get from the neighboring neurons connected to it. A number of methods are available in which information can be processed by a neuron and various types of ways to connect the neurons to each other. For the connection of neurons (with the help of using specific method) different types of neural network structures are constructed by using different type of processing elements. Pattern recognition, Control and signal processing, and many other types of applications in a variety of neural network structures were developed.

There are a significant number of Pattern-Recognition algorithms used for e-nose data processing and continuous developments are being made in this regard. The basic requirement for neural networks is their ability to follow the brain's pattern-recognition methods. The wide success of neural networks can be attributed to some key factors.

Neural networks, with their significant capability to derive meaning from complicated or indefinite data, can be used to remove patterns and detect development that are too difficult to be noticed by either humans or computer techniques. A trained neural network can work as an "expert" in the class of information it has been given for analysis. This skill can be used to provide projections given new situations of interest and to answer "what if" questions.

Self-Organizing Map (SOM)

Unsupervised training is the methods in which the networks learn to establish their own classifications of the training data without any external help. A self-organizing map is a special type of neural network for clustering purposes projected by Kohonen [12]. The main aim of the self-organizing map (SOM) is to transform a received signal pattern of arbitrary dimension into a two-dimensional discrete map, and to do this transformation adaptively in a topological ordered manner. These remarkable benefits proved it to be exceptionally successful for data visualization applications. The connection between the input data as measured

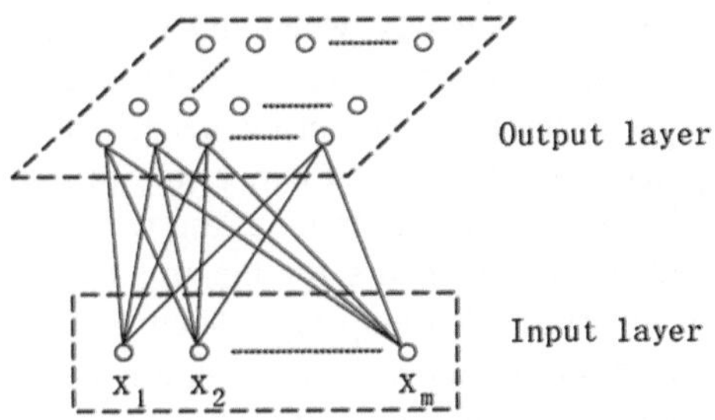

Fig. 7.23 the basic structure of SOM

in the input data space is preserved as faithfully as possible within the representation space. Thus, the similarity of the input data is reproduced to a very large extent in terms of geographical area within the representation space.

A single computational layer of neurons organized in rows and columns is arranged in feed forward structure. In the input layer all the source units are fully connected to each neuron. The basic structure of SOM is shown in Fig. 7.23.

For data clustering, the embedded competition paradigm is done by imposing neighborhood limitation on the output unit, such that a certain topological property in the input data is reflected in the outputs unit weights. SOM algorithm only require sensors output. SOM is a network created by N neurons arranged as the nodes of a planar grid.

The SOM Algorithm:

The aim is to study a *feature map* from the spatially *continuous input space*, in which input vectors live, to the low dimensional spatially *discrete output space,* which is formed by arranging the computational neurons into a grid.

The stages of the SOM algorithm that achieves this can be summarized as follows:

1. Initialization—Choose random values for the primary weight vectors w_j.
2. Sampling—Draw a sample training input vector x from the input space.
3. *Matching*—Find the winning neuron I(x)(best matching unit) with weight vector closest to input vector, i.e., the minimum value of $\boldsymbol{d_j(x)} = \sum_{i=1}^{D}(\boldsymbol{x_i} - \boldsymbol{w_{ji}})^2$
4. Updating—Apply the weight update equation $\Delta w_{ji} = \boldsymbol{\eta(t)T_{j,I(x)}(t)(x_i} - \boldsymbol{w_{ji}})$ Where $\boldsymbol{T_{i,I(x)}(t)}$ = Gaussian neighborhood and $\eta(t)$ is the learning rate.
5. Continuation—keep returning to step 2 until the feature map stops changing.(enough iterations for convergence)

The SOM is related to the category of competitive learning methods with unsupervised learning rules. It basically carryout analysis of a topology preserving projection of the data space onto a regular two-dimensional space where similar samples are located together. On the other hand, the basic self-organizing map has poor tracking capabilities when it is used with changing probability density of data. With the complete knowledge of odors stored in a single self-organizing map has the disadvantage of moving all the codebook vectors toward the new input probability distribution. This means that when the same odor is accessible in the extended time, all codebooks are moved toward the cluster that represents the current input, and even to others associated to different odors. Thus, other codebooks (related to different odor classes) become closer to the current one, destroying the historic knowledge base and making the system unstable for the task at hand. Other disadvantages of this technique is, the SOM that uses fixed network architecture in terms of number and arrangement of neural processing elements, has to be defined prior to training.

This problem has been solved by using an architecture based on multiple SOMs, each associated to a single odor to be recognized [13]. This network adjusts itself to new changes of the input probability distribution by means of repetitive self-training processes. Once each map self-organizes its codebook vectors, it refines them by using a learning vector quantization algorithm in order to reduce the high uncertainty accumulated at the borders of two or more different clusters. The self-training processes are carried out in an autonomous fashion during the testing phase to track odor patterns with changing statistical distributions [14].

For the case of mostly unknown input data characteristics it is available as insignificant to determine the network architecture that allows for satisfying results. During the unsupervised training process there is worth to consider a neural network models that determine the number and arrangement of units.

Cluster Analysis

The basic theory with these methods is that measurements made for related samples be likely to be similar. For similar samples the distance between samples is smaller compared to unrelated samples. It is a data reduction tool that creates subgroups that are more manageable than individual datum.

In cluster analysis there is no previous information about which elements belong to which clusters. The grouping or clusters are defined through an analysis of the data. For organizing observed data cluster analysis is an exploratory data analysis tool. Figure 7.24 illustrates the steps involved in the cluster analysis.

There are unsupervised cluster methods:

- Univariate clustering:
  It calculates individual variables and groups samples into homogeneous classes. Univariate clustering is useful to provide a proportional basis for exploring the data. Performing univariate study is a common way of exploring the data at hand, before turning to more complex analysis, such as multivariate analysis. The univariate analysis is applied on observable attributes.
- Hierarchical cluster analysis:
  In this analysis, reduction of multiple variables of a sample is done to a single 'distance' value. Rank and link samples are based on relative distances. For finding relatively homogeneous clusters of cases based on measured characteristics main statistical method is hierarchical cluster analysis. In this process, it starts with the each case as an element cluster, there are many clusters and these clusters are combines sequentially and it reduces the number of clusters at each

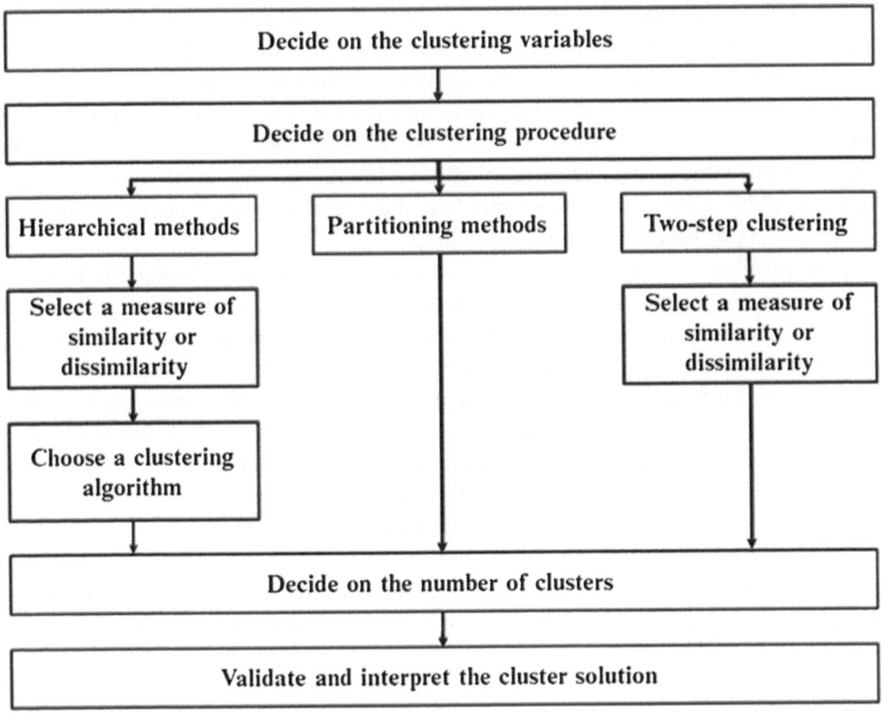

Fig. 7.24 Cluster analysis steps [15]

step until only one cluster is left. The clustering method uses the differences or distances between objects when structure the clusters. A dendrogram is the process represented on a diagram during execution of a hierarchical cluster analysis. This diagram demonstrates which clusters have been joined at each stage of the analysis and the distance between clusters at the time of joining.

- *k*-mean clustering:

k-mean clustering is the grouping of samples into a set number of classes and it uses all variables to determine relative distances. This method of clustering is dissimilar from the hierarchical clustering, which are applied when there is no prior knowledge of how many clusters there may be or what they are characterized by, so it is also called as non-hierarchical cluster analysis. *K*-means clustering is used when proposition concerning the number of clusters in cases or variables. The input required by the computer is to form exactly desired clusters that are to be as distinct as possible. The *k*-means clustering algorithm addressed this type of research question.

In general, the *k*-means method will produce the exact *k* different clusters demanded of greatest possible distinction. Frequently, both the hierarchical and the *k*-means techniques are used successively.

Cluster analysis has no method to distinguish between relevant and irrelevant variables. Therefore, the choice of variables included in a cluster analysis must be underpinned by theoretical considerations. This is very important because the clusters formed can be dependent on the variables included.

References

1. "Instrumentation engineering handbook", National nuclear security administration (NEVADA), Dec (2005)
2. H.W. Shin, A hybrid electronic nose system for monitoring the quality of potable water. Ph.D. Thesis, University of Warwick, School of Engineering, Oct (1999)
3. Ricardo Gutierrez-Osunae et al., A method for evaluating data-preprocessing techniques for odour classification with an array of gas sensors. IEEE Trans. Syst. Man Cybern. Part B: Cybern. **29**(5), 626–632 (1999)
4. P.C. Jurs, G.A. Bakken, H.E. McClelland, Computational methods for the analysis of chemical sensor array data from volatile analytes. Chem. Rev. **100**, 2649–2678 (2000)
5. M. Otto, in *Chemo-metrics*: *statistics and computer application in analytical chemistry*, (Wiley-VCH, NY, 1999), pp , pp. 119–174
6. S.M. Scott, D. James, Z. Ali, Data analysis for electronic nose systems. Microchim. Acta **156**, 183–207 (2007)
7. E. Llobet, E.L. Hines, J.W. Gardner, P.N. Bartlett, T.T. Mottram, Fuzzy ARTMAP based electronic nose data analysis. Sensors and Actuators B **61**, 183–190 (1999)
8. M.J. Adams, *Chemometrics in analytical spectroscopy* (Royal Society of Chemistry, Cambridge, 1995), pp. 70–79
9. J.W. Gardner, P.N. Barlett, *Electronic noses: principles and applications*, (Oxford University Press, Oxford, 1999)
10. L.I. Smith, A tutorial on principal components analysis, http://www.cs.otago.ac.nz/cosc453/student_tutorials/principal_components.pdf (2002)
11. J. Roden, D. Trout, B. King, A Tutorial on PCA Interpretation using Comp. Clust http://woldlab.caltech.edu/compclust/pca_interpretation_tutorial.pdf (2005)
12. T. Kohonen, Self organized formulation of topologically correct feature maps. Biol. Cybermatics **43**, 59–69 (1982)
13. C. Distante, P. Siciliano, K.C. Persuad, Dynamic cluster recognition with multiple self-organizing maps. Pattern Anal. Appl. **5**, 306–315 (2002)
14. M. Zuppa et al., Drift counteraction with multiple self-organizing maps for an electronic nose. Sens. Actuators B **98**, 305–317 (2004)
15. E. Mooi, M. Sarstedt, *A concise guide to market research: the process, data and methods using IBM SPSS Statistics*. (Springer, Berlin, 2011), ISBN:- 978-3-642-12540-9.(Chapter 9)

Chapter 8
Applications of Machine Olfaction

8.1 Introduction

Electronic-nose devices have gained huge attention in the field of sensor technology during the past 25 years, mainly the discovery of plentiful applications derived from research in varied fields of applied sciences.

In recent years, a growing worldwide awareness due to bionics and artificial intelligence has played a significant role in many features of human activity. The medical and microbiology fields are no exceptions; new socio-economical features and the need for developing a global community demands the development and application of new intelligent diagnostics and therapeutics near patient or home-based devices to control diseases efficiently.

Based on different detection principles and mechanisms, the development of many new E-nose sensor types and arrays is closely correlated to the expansion of new applications. Precise analysis of E-nose depends on many factors such as sensor types, pattern recognition algorithm, software development, and environment. The sensor is one of the important factors for successful and accurate odor measurement using electronic nose (E-nose). Hence, the discovery and design of many new E-nose sensory types and arrays based on various principles and mechanisms provide vast uses of E-nose in new applications [1]. These inventions and modifications in E-nose provide a number of benefits to mankind, including commercial, and help to improve human society lifestyles.

Following are the list of industries where E-nose is used:

- The agricultural industry,
- Biomedical industry,
- Cosmetics industry,
- Environmental monitoring,
- Food analysis,
- Manufacturing industry,
- Military application,
- Pharmaceutical industry, and
- Quality control in laboratory.

H. K. Patel, *The Electronic Nose: Artificial Olfaction Technology*,
Biological and Medical Physics, Biomedical Engineering,
DOI: 10.1007/978-81-322-1548-6_8,

The sense of smell plays a key role in human development and biosocial interactions. In the development of many commercial industries, managing aroma properties to improve product appeal, quality, and consistency by individual brands classification is done by users having unique scents. Hence it is an essential requirement of the olfactory sense. From wines and cuisine, perfumes and colognes added to personal healthcare products, and scents applied to product packaging that signify the importance of aroma qualities in industrial manufacturing and commercial trade, are examples of E-nose implementation in various types of industries [2].

Sensations find out the human being. It also might be true that olfactory experiences mediate the first contact of the human being with the world, through the mother's smell, which establishes his survival and represents the beginning of his understanding.

Electronic nose is generally used for different applications in the food and beverage industries: Identification, quantification, and quality control. In every application, the principal goal is that the instrument should distinguish different properties of different samples. An individual's properties could be qualities, origins, defects, and concentration of pollutants. One of the uses of this instrument is that when applied on the analyzed products, the same structure as the one induced by a human sensory panel allows instrumental measurement of sensory properties.

In the following pages of this chapter, various individual examples of E-nose applications in each industry (i.e., automobile, food, packaging, cosmetic, drug, etc.) and product areas (monitoring, analysis, classification, etc.) are discussed and they are summarized in Table 8.1 [2].

8.1.1 Food Industry

The food industry is the biggest market for E-nose. Electronic nose utilization in this field includes Inspection of the nature and quality of ingredients, supervision of the manufacturing process, and all processes related to the shelf life of the product. Quality assurance systems are required in the food processing industry.

Generally, qualitative assessment of food spoilage/worth is made by human sensory panels that assess air samples and discriminate which food products are good or offensive. In some cases, E-noses can be used to supplement or replace panels of human experts. In other cases, E-noses can be used to reduce the amount of analytical chemistry performed in food production, particularly when qualitative results will carry out.

(a) Milk and dairy products

Electronic nose mainly focuses on the categorization of Parmesan cheeses with rates of maturity for the different types. The E-nose effectively differentiates the types and characterizes cheese using polypyrrole semiconductor sensors [3]. Milk spoilage due to microbial/bacterial contamination can be identified from the presence of acetaldehyde, acetic acid, and ethanol [4, 5] in headspace of the

Table 8.1 Industry-based applications for electronic noses [2]

Industry area	Application	Specific use types and examples
Agriculture	Crop protection, harvest timing and storage, meat, seafood, and fish products, plant production	Homeland security, safe food supply
		Crop ripeness, preservation treatments, freshness, contamination, spoilage
	Pre- and post-harvest diseases	Cultivar selection, variety characteristics, plant disease diagnoses, pest identification, detect non-indigenous pests of food crops
Airline transportation	Public safety and welfare	Explosive and flammable materials detection
	Passenger and personnel security	
Cosmetics	Personal application products	Product enhancement, consumer appeal
	Fragrance additives	Perfume and cologne development
Environmental	Air and water quality monitoring	Pollution detection, effluents, toxic spills malodor emissions, toxic/hazardous gases, control of point-source pollution releases
	Indoor air quality control	
	Pollution abatement regulations	
Food and beverage	Consumer fraud prevention	Ingredient confirmation, content standards, brand recognition, product consistency, marketable condition, spoilage, shelf life, off-flavors, product variety assessments
	Quality control assessments	
	Ripeness, food contamination	
	Taste, smell characteristics	
Manufacturing	Processing controls, product uniformity, safety, security, work conditions	Product characteristics and consistency
		Aroma and flavor characteristics
		Fire alarms, toxic gas leak detection
Medical and clinical	Pathogen identification	Patient treatment selection, prognoses
	Pathogen or disease detection	Disease diagnoses, metabolic disorders
	Physiological conditions	Nutritional status, organ failures
Military	Personnel and population security civilian and military Safety	Biological and chemical weapons
		Explosive materials detection
Pharmaceutical	Contamination, product purity	Formulation consistency and uniformity
	Variations in product mixtures	Quality control of drug purity
Regulatory	Consumer protection	Product safety, hazardous characteristics air, water, and soil contamination tests
	Environmental protection	
Scientific research	Botany, ecological studies	Chemotaxonomy, ecosystem functions
	Engineering, material properties	Machine design, chemical processes
	Microbiology, pathology	Microbe and metabolite identifications

milk, which are absent in raw milk before spoilage. Metal oxide semiconductor-based E-nose sensors are used for real-time quality analysis of raw milk. The sensors are empaneled and calibrated toward various concentrations of volatile organic compounds (VOCs) which are liable for off-flavors (ethanol, trimethylamine, acetaldehyde, dimethylsulfide, acetic acid, etc.) in milk produced due to microbial contagion, chemical reactions, and genetic effects in cow.

An article published by S. Ampuero et al., which reviews E-nose *applications in the milk industry* : i.e., examples of the analysis of Swiss and Cheddar cheese aroma; the measurement of the ripening of Pecorino Toscano cheese (ewe's); the detection of mold in Parmesan cheese; the categorization of milk by trademark, by fat level, and by preservation process; the classification and the quantification of off-flavors in milk; the evaluation of Maillard reactions during heating processes in block-milk; and the identification of single strains of disinfectant-resistant bacteria in mixed cultures in milk [5].

(b) Meat products

The major important factors in meat and meat products are microbiological safety, shelf life of products, and temperature. E-nose measurement systems can analyze changes in the headspace volatiles of aerobically packed meat. The E-nose has the benefits of being rapid, non-destructive, and non-contact instrumental testing [6]. Mahdi Ghasemi-Varnamkhasti et al. explain the applications of E-nose systems for meat quality assessment, where fast detection methods are necessary for proper product management. The outcome suggests the option of using this new technology in meat handling [7].

(c) Fish and seafood products:

Freshness is the most important factor for fish quality. Electronic nose (E-nose) can be used to identify odor variety, aroma intensity, and degree of freshness. GholamHosseini, H. et al. proposed artificial neural networks (ANNs) to design an intelligent system for measuring the freshness of spoiling fish. They proposed 32 sensory arrays-based portable E-noses [8].

(d) Fruity odors classification:

Kea-Tiong Tang et al. developed a prototype of a portable E-nose with the use of eight sensor arrays which are commercially available sensors and this E-nose is able to detect and identify the fragrance of fruity odors [9].

(e) Classification of beverages:

Mazlina et al. designed and developed an E-nose sensor system using 14 metal oxide semiconductor gas sensors from Figaro Sensor, to classify different beverages and working as good analytical instruments with good characteristics (repeatability, reproducibility, and discriminative ability). It is also able to produce different patterns for different samples [10].

(f) Cocoa beans quality assessment:

Olunloyo et al. designed and developed a prototype E-nose sensor for monitoring the quality of cocoa beans. They employed in this work a sensing element manufactured by Figaro TGS (Figaro website: www.figaro.com) gas sensor, which includes a Tin Oxide (SnO_2) semiconductor, with low conductivity in clean air. The compounds dependable for the cocoa aroma are methylpyrazines, which are nitrogen heterocyclic substances, known as volatile compounds in the aroma of five types of roasted and unroasted (raw) cocoa beans as isovaleraldehyde, isobutyraldehyde, propionaldehyde, methyl alcohol, acetaldehyde, methyl acetate, n-butyraldehyde, and diacetyl. They used PCA method for baseline correction of data and normalization of data [11].

(g) Honey Characterization:

Honeys can be distinguished according to their botanical origin on the basis of volatile compound outlines obtained by E-nose. The differences caused by the honey types (obtained differences originating from the geographic samples of honey) were counteracted by Frane et al. [12]

(h) Aroma Detection of Coffee:

Schedule analysis is frequently performed on coffee aromatic extracts to evaluate the effectiveness of the extraction methods used in interpretation in a quality of coffee aroma. A good extraction method is expected to provide an extract with sensory characteristics very close to the aroma of ground coffee beans prior to extraction.

(i) Quality of Tea:

There are around 700 volatile organic compounds (VOCs) in black tea and Linalool, Geraniol, Trans2-hexen-1-ol, and Methyl Salicylate which are present in reasonably high concentrations; these concentrations increase the good quality of black tea. As there are quite a large number of volatile compounds present in tea, a sensor array, comprising a group of sensors, is the best option for assessment of tea quality [13].

(j) Wine Fermentation:

Fermentation in wine is the process whereby yeast transforms sugar into carbon dioxide and ethyl alcohol (ethanol). The proposed E-nose (Wi-Nose) was successfully designed to examine the fermentation stages in wine. With the help of neural networks, 100 % accuracy is possible to be realized. The product quality is maximum consumer satisfaction, which is better than the competition, and can be marketed optimally at a price slightly lower than the competitor price [14].

8.1.2 Medical Industry

In various cases, infection with microorganisms generates a change in the way a person smells, which can be especially perceptible on the breath, in urine, or stools; such changes are used to assist in the diagnosis of diseases and in some countries, smelling the patient or the body fluids of patients was, and still is, an important tool in diagnosis.

Explore the potential applications of using qualitative volatile fingerprints (E-nose) for early detection and diagnosis of diseases such as dermatophytosis, ventilator associated pneumonia, cancer, renal analysis, and asthma detection. There is a necessity for preventive medical checkups to diagnose disease early, to speedup the healing process, to increase the rate of complete recovery, and accordingly, to save money for the healthcare system [15]. Table 8.2 includes a list of molecular biomarker VOCs of particular human diseases and disorders and Table 8.3 presents existing E-nose uses in hospitals and universities for healthcare/medicines [15].

Table 8.2 Molecular biomarker VOCs of particular human diseases and disorders [15]

Disease/disorder	Volatile chemical biomarkers
Allograft rejection	Carbonyl sulfide
Breast cancer	C4-C20 alkanes
Cholera	p-menth-1-en-8-ol, dimethyl disulphide
Chronic hepatitis	Methyl-mercaptan, dimethyl sulfide
Cirrhosis	Dimethyl sulfide, mercaptans
Cystic fibrosis	Leukotriene B4, interleukin-6, carbonyl sulfide, alkanes
Diabetes	Acetone, ethanol, methyl nitrate
Halitosis	Methanethiol, Hydrogen sulfide, methyl mercaptan, dimethyl sulfide
Hepatic encephalopathy	3-methylbutanal
Histidinemia	2-imidazolepyruvic acid, 2-imidazolelactic acid, 2-imidazoleacetic acid
Liver cancer	Hexanal, 1-octen-3-ol, octane
Lung cancer	Alkanes, ketones, specific aromatic hydrocarbons (benzene derivatives)
Maple syrup disease	2-oxoisocaproic acid
Necrotizing enterocolitis	2-Ethyl-1-hexanol
Oxidative stress	8-isoprostane
Periodontal disease	Pyridine, picolines
Phenylketonuria	Phenylpyruvic acid, phenyllactic acid, phenylacetic acid
Schizophrenia	Pentane, carbon disulfide
Tyrosinemia	p-hydroxyphenylpyruvic acid
Trimethylaminuria	Trimethylamine
Uremia	Dimethylamine, trimethylamine

Table 8.3 Electronic nose uses in hospitals and universities for healthcare/medicines [15]

Country	Hospital, university or research facility	E-nose utilized	Application
USA	University of Pennsylvania	Experimental model	Distinguish cerebrospinal fluid
USA	Merck Research Laboratories	FOX 4000	Flavor analysis for Drug formulation
United Kingdom	Birmingham Heartlands hospital	Cyranose 320	Identify *Staphalococcus*
Germany	University of Applied Sciences	DE101	Detect renal dysfunction
USA	Cleveland clinic	Sensor array (Metabolomx manufacturer company)	Diagnose lung cancer
Belgium	University of Antwerp	PEN2	Clinical diagnoses of bacteria
United Kingdom	South Manchester University Hospital	Experimental model	Burn and wound Infection types
USA	University of Pennsylvania	–	Diagnosis of diseases via breath
USA	California Institute of Technology	JPL Enose	Detect & differentiate brain cancers
Australia	Prince Charles Hospital	–	Detect chronic lung disease
Netherlands	Amsterdam Academic Medical Center	Cyranose 320	Discriminate inflammation Airway diseases
Italy	Catholic University	Experimental model	Asthma detection
Tanzania	National Institute of Medical Research	Bloodhound EN	Diagnosis of tuberculosis
United Kingdom	Gloucestershire Royal Hospital	NST 3320	Diagnosis of ventilator associated pneumonia
Australia	St Vincent's Clinical School, New South Wales	CyraNose 32	Asbestos-related disease of mesothelioma

8.1.2.1 Bacteria Identification in Blood Culture Samples/Urine Analysis/ENT Bacteria/Upper Respiratory

Discovery of diagnostic methods that give fast detection of the presence and type of bacteria and so allowing proper antibiotic treatment can provide an increased benefit to the patient as well as help reduce cost to the health care system. The use of E-noses is to differentiate among various bacteria regularly found in blood cultures [16], upper respiratory disease identification [17], ventilator-associated pneumonia [18], and urine analysis [19, 20].

The E-nose can distinguish among various common bacterial pathogens of the upper respiratory area, including Staphylococcus aureus, Streptococcus pneumoniae, Haemophilus influenza, and Pseudomonas aeruginosa [17]. This is an important application of electronic olfaction that could considerably improve the current methodologies and be successfully used in clinical settings.

8.1.2.2 Ventilator Associated Pneumonia

Ventilator associated pneumonia (VAP) has a high rate of mortality and increased incidence in critically ill individuals. Humphreys et al. explored the potential of volatile fingerprints produced by different microorganisms not only *in vitro* but also within clinical samples using an E-nose in perceptive between these groups. They also tried to associate the microbiology culture results with the E-nose responses for clinical samples [18]. The use of an E-nose is shown in recent studies on tuberculosis for perception between various mycobacterial isolates in humans [21].

8.1.2.3 Monitoring Haemodialysis

The study was done to analyze the potential for an "E-nose" as a basic monitoring tool for haemodialysis. Blood samples were analyzed using an E-nose, which comprised an array of 14 conducting polymer sensors, and these E-nose results were compared to traditional biochemistry. E-nose demonstrated the ability to distinguish predialysis blood from postdialysis blood independently, together with an appropriate classification model, suitable for online monitoring [22]. The study also suggests that E-nose technology might be a useful tool in discriminating predialysis from postdialysis blood as well as control blood. Simultaneously, with an appropriate classification model, it might be possible to build an online monitoring system for the management of renal failure. Also it should be possible to improve and modify currently used sensor arrays toward specific volatile markers or marker groups, which would simplify the optimization of such applications.

8.1.2.4 Lung Cancer Detection

The exhaled breath of patients with lung cancer has different unique characteristics that can be identified with an E-nose made with the use of polymer sensory array. The results provide available options to the concept of using the E-nose for administration and detecting lung cancer [23].

Uncontrolled cell growth starting in the lungs is defined as lung cancer. The establishment of this excessive growth is called hyperplasia. Hyperplasias begin in the bronchial tubes or alveoli. The bronchial tubes are in direct contact with air and its impurities. Blatt et al. concluded remarkable detection results of lung cancer diagnosis using MOS sensors [24].

In another research, Guo et al. conducted a study on diagnosis of diabetes, renal disease, and airway inflammation using MOS sensor array, with the result that breath analysis can be employed for early diagnosis of many diseases [25].

8.1.2.5 Diagnosis of TB

Arend et al. explored the prospect of two different E-noses (EN, code named "Rob" and "Walter") to differentiate between sputum headspace samples from TB-patients and non-TB patients [26]. With the help of E-nose a sample test (if the patient is on field) is done and the early diagnosis of tuberculosis followed by appropriate treatment will reduce the trouble of tuberculosis.

8.1.2.6 Diagnostic Tools: Otolaryngology

Electronic nose technology and successful clinical application are particularly new methods for the fast near-patient diagnosis of disease. There is real clinical need for new diagnostic tools in otolaryngology. E-nose technology provides noteworthy potential for rapid, noninvasive, point of care diagnosis of disease in otolaryngology [27].

8.1.3 Agriculture

The applications of E-noses in the environment and agriculture fields are in general meant to replace the slow and difficult laboratory analysis with fast and easy in-field E-nose analysis. The frequent applications of E-noses in agriculture are to monitor food quality and production processes, detect crop diseases, and identify insect infestations.

8.1.3.1 Plant Pathology

The E-nose technology is new and in its infant stage for application in plant pathology.

Electronic nose is a rapid, sensitive, specific, and easy to use technique utilized for detection and identification of plant pathology, which includes identification of plant pathogenic bacteria in clinics and laboratories.

The discrimination of seven species of plant pathogenic bacteria (*Acidovorax avenae* subsp. *citrulli*, *Agrobacterium tumefaciens*, *Clavibacter michiganensis* subsp. *michiganensis*, *Erwinia amylovora*, *Pseudomonas syringae* pv. *tomato*, *Ralstonia solanacearum*, and *Xanthomonas campestris* pv. *vesicatoria*) by measuring the volatile compounds produced from pure cultures has been performed using an e- nose and discriminant function analysis [28].

A. C. Bastos et al. examined in their study the potential of using an array of nonspecific conducting polymer sensors for observants between soil types and different soil environmental conditions and treatments. Results indicate that qualitative soil fingerprint analyses using E-nose technology can be employed as a rapid, sensitive, and noninvasive tool for characterizing soil status, identify changes in soil conditions, and monitor processes such as organic matter degradation [29]. In Table 8.4 offensive agricultural by-products are listed with human detection threshold level and recognition level (measured in parts per million (ppm) in dry air at standard temperature and pressure) [30].

8.1.4 Environmental

Environmental applications of E-noses include analysis of fuel mixtures, detection of oil leaks, testing ground water for odors, and identification of household odors, identification of toxic wastes, air quality monitoring, and monitoring factory emissions. For effective detection and monitoring, the beginning activity of *Streptomyces* in water at different stages of differentiation, as well as to differentiate between different species based on their volatile production patterns, could be successfully made by an E-nose [31].

It also suggests that it could be used for monitoring geosmin production in water and possibly set threshold odor levels, as a routine task for specific water-screening purposes. E-nose technology also offers prospects for replacing existing techniques for environmental applications, in a quick and highly reproducible way.

Bourgeois and Stuetz [32] reported the use of conducting polymer sensor arrays to analyze wastewater samples and monitor changes in wastewater quality, and provide a simple noninvasive technique for online monitoring of wastewater.

An illness associated with poor ventilation has determined an increasing consideration toward indoor air quality monitoring. Miniaturized, low-cost E-nose developed by using metal oxide sensors integrated within climate control unit and signal processing techniques were developed [33].

Table 8.4 Offensive agricultural by-products (threshold level for human detection and recognition) [30]

Chemical odorant	Formula	Characteristic odor	Detection	Recognition
Acetaldehyde	CH_3CHO	Pungent, fruity	1.6×10^{-2}	2.1×10^{-1}
Allyl mercaptan	CH_2CHCH_2SH	Strong garlic, coffee		
Ammonia	NH_3	Sharp, pungent		
Amyl mercaptan	$CH_3(CH_2)4SH$	Putrid		
Benzyl mercaptan	$C_6H_5CH_2SH$	Strong		
Butylamine	$C_2H_5(CH_2)2\ NH_2$	Ammonia-like, sour		2.4×10^{-1}
Cadaverine	$H_2N(CH_2)5\ NH_2$	Putrid, decaying flesh		
Chlorophenol	ClC_6H_5O	Phenolic, medical		
Crotyl mercaptan	$CH_3CH = CHCH_2SH$	Skunk-like	7.7×10^{-3}	
Dibutylamine	$(C_4H_9)_2NH$	Fishy		
Disopropylamine	$(C_3H_7)_2NH$	Fishy		8.5×10^{-2}
Dimethyamine	$(CH_3)_2NH$	Putrid, fishy		4.7×10^{-2}
Dimethylsulfide	$(CH_3)_2S$	Decayed vegetables		1.0×10^{-3}
Diphenylsulfide	$(C_6H_5)_2S$	Unpleasant		2.1×10^{-3}
Ethylamine	$C_2H_5NH_2$	Ammonia-like		8.3×10^{-1}
Ethyl mercaptan	C_2H_5SH	Decayed cabbage	2.6×10^{-3}	1.0×10^{-3}
Hydrogen sulfide	H_2S	Rotten eggs		4.7×10^{-3}
Indole	C_2H_6NH	Nauseating, fecal		
Methylamine	CH_3NH_2	Putrid, fishy		2.1×10^{-2}
Methyl mercaptan	CH_3SH	Decayed cabbage		2.1×10^{-3}
Propyl mercaptan	$CH_3(CH_2)_2SH$	Unpleasant	2.4×10^{-2}	
Putrescine	$NH_2(CH_2)_4NH_2$	Putrid, nauseating		
Pyridine	C_6H_5N	Disagreeable, irritating		
Skatole	C_9H_9N	Nauseating, fecal	2.2×10^{-1}	4.7×10^{-1}
Sulfur dioxide	SO_2	Pungent, irritating		
Tert-butyl mercaptan	$(CH_3)_3CSH$	Unpleasant, skunk		
Thiocresol	$CH_3C_6H_4SH$	Rancid, skunk	1.4×10^{-2}	
Thiophenol	C_6H_5SH	Putrid, garlic-like	1.4×10^{-2}	2.8×10^{-1}
Triethylamine	C_2H_5OH	Ammonia-like, fishy		

The odorous element of the atmospheric emissions is not, in most cases, a threat to public health, since it is due to substances whose concentration level is far below the threshold limit value set by health authorities. An E-nose intended for evaluation of odor annoyance associated with atmospheric emissions was presented by Fabio et al. The E-nose included an array of conducting polymer sensors for odor detection, and a fuzzy algorithm for data analysis [34].

The E-nose measurement system was designed based on an array of metal oxide sensors used to classify three applications which include monitoring of wastewater volatiles, detection of dry rot infections in buildings, and smart fire detection systems [35]. Quality control of environmental measurement can be achieved with speed, precision, and accuracy. E-nose developed with the use of SAW sensor analyze vapor in air, water, and solid [36].

8.1.5 Cosmetic Industry

From skin, body, and hair care products including cosmetics, the consumer tendency is toward using naturally resultant products. An E-nose using CP sensors was developed and used for analysis of off-skin odors [37].

8.1.6 Military Application

An E-nose using fast gas chromatography (GC) and the surface acoustic wave (SAW) resonator detector for the detection of explosives is described to detect VOC from soil and groundwater such as explosives [38].

M. C. Burl et al. performed a laboratory experiment to study the rapid detection of very low concentrations of dinitrotoluene (DNT) using a conducting polymer composite-based E-nose. DNT is an impurity that results during the production of military grade TNT. DNT has been shown in the soil beyond covered land mines; so detection of DNT has gained intense attention as a possible tool for demining [39].

8.1.7 Space Application

A skill to observe the basics of breathing air becomes complicated when trying the same for maintaining human health in a closed environment. NASA is developing an E-nose for monitoring capability for use in closed environments in which air is recycled, such as in space shuttles, in space stations, and in planned space human environments. A miniature E-nose has been designed and developed at the Jet Propulsion Laboratory (JPL), Pasadena, CA, which was able to detect, identify, and quantify 10 toxins and relative humidity changes. The E-nose sensing array includes 32 sensing films made from polymer carbon-black composites [40].

8.1.8 Pharmaceutical Industry

Flavors are used in pharmaceutical oral solutions and oral suspensions to cover drug bitterness and to make the formulation more edible. A new approach using a metal oxide sensor-based electronic-nose system for headspace analysis was explored to obtain consistent qualitative and quantitative analysis of flavors in a pharmaceutical formulation. The utilization of the E-nose technique was demonstrated to qualitatively differentiate among six common flavoring agents (raspberry, red berry, strawberry, pineapple, orange, and cherry) in formulations. It can also be used for packaging selection for the formulation to ensure the flavor shelf life [41].

8.1.9 Case Study

8.1.9.1 Human Breath

Breakthrough of VOCs found in breath, guide to its use in the diagnosis of diseases, disorder, and metabolic learning. Breath can be analyzed without analytical instruments, i.e., exhaled breath VOC chemicals defined the symptoms of diabetes, asthma, COPD, bronchiectasis, and lung diseases related to the respiratory system. Breath analysis is an easy process, which does not have pain or embarrassment related with blood and urine analysis. However, breath analysis has the problem that it contains more water content in preconcentration sampling and detection of substances. VOCs are acquiesed of core metabolic processes while inorganic molecules are related to health conditions and can mimic a probable disease of the individual or a recent exposure to a drug or an environmental pollutant. Therefore, the abnormality in the concentration of certain trace gases, so-called biomarkers, could provide clues to diagnose corresponding diseases.

The exhaled human breath is a mixture of N_2, O_2, CO_2, H_2O, inert gases, and thousands of other trace gases. These gases include inorganic molecules such as NO, NH_3, or CO, and volatile organic compounds (VOCs) such as acetone, methanol, or isoprene, with concentrations ranging from ppb to ppm as shown in Table 8.5. Example of taking breath sampling from human breath using sampling system is shown in Fig. 8.1.

There are various techniques by which breath can be analyzed as in GC and its types are (1) Mass Spectroscopy (MS), (2) Flame-Ionized Detection (GC-FID), (3) Ion Mobility Spectroscopy (IMS), and (4) Electrolyzer powdered flame ionization. Table 8.6 lists breath molecules origin and Table 8.7 summarizes biomarkers and their respective physiological concentration ranges in the human breath.

Table 8.5 Exhaled breath molecules concentration

Concentration	Molecule
Percentage	Oxygen, water, carbon dioxide
Parts-per-million	Acetone, carbon monoxide, methane, hydrogen, isoprene, benzene methanol
Parts-per-billion	Formaldehyde, acetaldehyde, 1-pentane, ethane, ethylene, other hydrocarbons, nitric oxide, carbon disulfide, methanol, carbonyl sulfide, methanethiol, ammonia, methylamine, dimethyl sulfide, benzene, naphthalene, benzothiazole, ethane, acetic aide

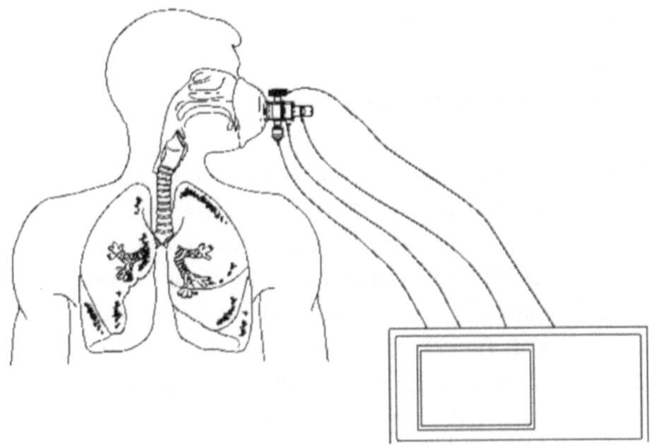

Fig. 8.1 Breath sampling system [52]

Table 8.6 Physiological origin of some breath molecules

Breath molecules	Physiological origins
Acetaldehyde	Ethanol metabolism
Acetone	Decarboxylation of acetoacetate
Ammonia	Protein metabolism
Carbon disulfide	Gut bacteria
Carbon monoxide	Production catalyzed by heme oxygenase
Carbon sulfide	Gut bacteria
Ethane	Lipid peroxidation
Ethanol	Gut bacteria
Ethylene	Lipid peroxidation
Hydrocarbons	Lipid peroxidation/metabolism
Hydrogen	Gut bacteria
Isoprene	Cholesterol biosynthesis
Methane	Gut bacteria
Methanethiol	Methionine metabolism
Methanol	Metabolism of fruit
methylamine	Protein metabolism
Nitric oxide	Production catalyzed by nitric oxide synthase
Pentane	Lipid peroxidation

8.1.10 Case Study 1: Practical Approach to Detect Asthma in Human Breath

8.1.10.1 The Problem

Measurement of volatile compounds and odors in breath using E-noses is the low-cost, real-time, qualitative, and portable technique to carry out reliable, reproducible, and objective measurement. To achieve a reliable, subjective, and analytically acceptable system it is required the awareness of the difference between

Table 8.7 Biomarkers and their respective physiological concentration ranges in the human breath [53]

Biomarkers	Physiological origin	Related diseases	Physiological ranges in human breath
Ethane	Lipid peroxidation	Oxidative stress	1–11 ppb
Pentane	Lipid peroxidation	Oxidative stress	Less than ethane
Isoprene	Cholesterol biosynthesis	Cholesterol metabolic disorder	55–121 ppb; 12–580 ppb;
Acetone	Decarboxylation of acetoacetate and acetyl-CoA	Diabetes mellitus, ketonemia	293–870 ppb; 1.2–1,880 ppb
Ethanol	Alcohol ingestion	Alcohol poisoning	27–153 ppb; 13–1,000 ppb
Methanol	Degradation of natural pectin from plants; ingestion	Methanol intoxication	160–2,000 ppb
NH_3	Metabolic product of amino acid deamination	Uremia, kidney impairment	422–2,389 ppb; 200–1,750 ppb
CO	Inhalation from Incomplete burning of carbon containing fuels, e.g. smoking	Lung diseases	<6 ppm
NO	L-arginine oxidation	Asthma, lung diseases	1–9 ppb, lower respiratory; 0.2–1 ppm upper respiratory; 1–30 ppm, nasal level

an artificial system and physiology system. The E-nose system developed to mimic the function of human noses is developed with more precision and is everlasting.

Research is ongoing on the use of E-noses to diagnose illness by smelling patient's breath with the option of installing tiny E-noses in phone receivers so that patients can simply breathe into the phone and wait for diagnosis. To detect subtle changes in body by odor which can help to indicate primary disease identification or disease present conditions, high technology sniffers—E-nose are used just by smelling the breath. Hence, there are a number of applications already developed and still developing with the use of E-nose as we already discussed in the application overview.

To measure the concentrations of specific biomarkers in exhaled human breath, different techniques/methods have been developed and can be classified into two groups:

- Spectrometry/spectroscopy-based techniques
- Chemical sensors

The E-nose system makes use of chemical sensors that are particularly sensitive to the biomarkers and compositions in human breath to trigger responses to a patient's breath sample. In contrast to the broad panel of nonspecific sensors used

Fig. 8.2 Tedlar bag

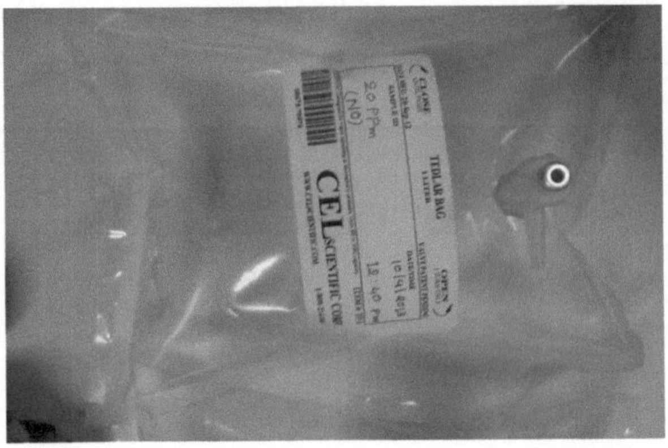

in commercial E-noses, the sensors of the proposed system were specifically selected for their responses to known components of human breath. The sample is injected into the system using a Tedlar bag (as shown in Fig. 8.2) to guarantee all samples are sampled under the same criterion. The chemical sensors sense the sample and accordingly form a kind of "odour print" that is typically associated with a given disease or condition. The "odour print" is then sent to the computer for signal processing and pattern recognition.

Patients have to give their breath samples in two ways depending on the requirement of biomarkers for disease diagnosis:(1) Depending on whether the condition under consideration typically exhibits its biomarkers in what are known as dead space air from the upper airway, or alveolar air from the lungs; and (2) Depending on the type of biomarkers and on the breath test tracks, dead space air may be either a necessity or a contaminant. Dead space air is required when the biomarkers are released into the airways, and thus into the dead-space air.

8.1.10.2 Methods

Nitric oxide is a type of inorganic gas with small molecules and simple structure (N=O). Nitric oxide is one of the least known biologically active messenger molecules. In spite of a momentary gas, which is rapidly oxidized into NO_2, nitrites, and nitrates by O_2, it is fairly stable at low concentrations; even O_2 is present [42]. NO, a free essential, is a gaseous signaling molecule which plays an important role in normal airways and in blood vessel quality parameter. NO is generated by NO synthesis in the normal human airway. Normal exhaled breath includes low PPB levels of NO, however, asthmatic patients, who seem to have increased inducible NO synthase in their airways prove to have a prominent concentration of FeNO [43].

Asthma is a persistent provocative disorder of the airways that generates airway hyper responsiveness, reversible airway obstruction, and symptoms such as wheezing, cough, and shortness of breath. A boost in exhaled NO is not specific for asthma, but an

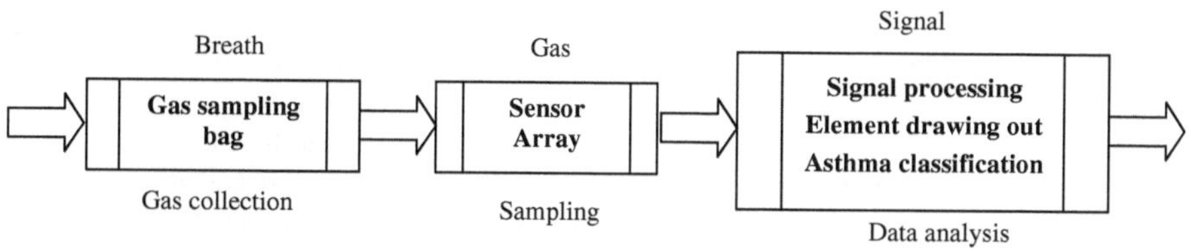

Fig. 8.3 Proposed asthma analysis system in breath

increased concentration may be useful in distinguish asthma from other causes, i.e., chronic cough [44]. Dupont et al. presented that >16 ppb of NO in the lower respiratory tract could be treated as a cutoff for asthma with a 90 % specificity and 90 % positive predictive value. During research they analyze the diagnostic value of exhaled NO capacity to differentiate between healthy persons with or without respiratory symptoms and patients with confirmed asthma [45]. Thus it implies that noninvasive measurement of exhaled NO is a good rival for asthma diagnosis.

This proposes a novel system that is special for breath analysis. We selected chemical sensors that are sensitive to the biomarkers and compositions in human breath, developed the system, and introduced the odor signal preprocessing and classification method. To evaluate the system performance, we captured breath samples from healthy persons and patients known to be afflicted with asthma and conducted experiments on medical treatment evaluation and disease level identification. The breath sample is passed over the sensor and the sensor output is stored in the PC through a software developed in LabVIEW as shown in Fig. 8.3. The sensor response of the patient is then compared with the database. Based on the patient database developed, the treatment plan is suggested for the specific level of asthma, if detected. If the sensor response for the sample is below the reference line, the person is declared non-asthmatic. The reference line is fixed through the breath sample of a healthy person.

Sensor Associated with Disease Condition

The chemical sensor capabilities can be identified with the system function and performance. Each sensor in the array has a unique "odour print," corresponding to the compounds. The majority of the compounds are VOCs, but some are inorganic compounds, such as ammonia, nitric oxide, carbon dioxide, and hydrogen. Table 8.8 summarizes the main disease biomarkers and compositions in human breath and the type of sensor required.

Sensor

In this work, we have utilized gas sensors, which are commercially available from alpha sense ltd. and Figaro. These gas sensors have been commonly identified as

Table 8.8 Compounds detected in human breath and sensors Required

Molecules in human breath	Requisite sensors	Disease condition
Acetone, isoprene, pentene, benzene, etc.	VOC sensor	Diabetes
Ammonia	NH_3	Renal function
Carbonyl sulphide, carbon disulphide	Sulphide sensor	Organ rejection, Liver function, schizophrenia
Carbon monoxide, carbon dioxide	CO and CO_2 sensor	Ulcers
Nitric oxide	NO sensor	Asthma
Hydrogen	H_2 sensor	

amperometric electrochemical gas sensors. Table 8.4 contains specifications of sensor (Alpha sense Ltd.) which used four electrode sensors as sensor array used for developing miniature sensor used in specific analysis of E-nose system (Table 8.9).

NOB4 sensor complies with the required characteristics when the sensor is tested in standard conditions as specified as follows:

Sensitivity	nA/ppm in 5 ppm NO	700–1300
Response time	t_{90} (s) from zero to 5 ppm NO	<80
Zero current	ppb equivalent in zero air at 20 °C	<55
Noise[a]	RMS noise (ppb equivalent)	<5
Limit of detection[a]	ppb equivalent	<3
Range	ppm no limit of performance warranty	20
Linearity	ppm error at full scale, linear at zero and 5 ppm NO	<±1
Over gas limit	Maximum ppm for stable response to gas pulse	50

Alpha sense toxic gas sensors are prepared using three electrochemical cells that operate in the amperometric mode. The output of this sensor current is linearly proportional to the partial volume of the toxic gas such as CO or H_2S. Figure 8.4 shows the schematic of sensor and Fig. 8.5 shows the structure of a toxic gas sensor [46]. Figure 8.6 shows the potentiometric circuit which will minimize the current transients by keeping the working electrode at a stable potential. While electrochemical toxic gas sensors are planned to operate in the stability area region to reduce sensitivity to small reference electrode-potential changes, one cannot conclude that small changes in the reference electrode potential can be ignored.

The Figaro sensors (TGS) are available for detection of exhaled breath for diagnosis of asthma in human. Deposition of a metal oxide semiconductor (MOX), i.e., SnO_2 and WO_3, as a thin film on inter digit electrodes and catalytic reactions of the metal oxide surface with the target gas molecules, temperature between 250 and 350 °C, the resistance between the electrodes is changed, and measured in the TGS sensors development. A simple circuit is required for performance is the advantage of MOX sensors (resistive) over the other types of gas sensors. Design

Table 8.9 Specifications: B4 (4-Electrode sensors) [46]

Miniature sensors	H_2S-B4	CO-B4	**NO-B4**	NO_2-B4	SO_2-B4	O_3-B4
Operating range (ppm/ % LEL)	0–25	0–50	**0–20**	0–20	0–20	<2
Sensitivity (nA/ppm)	1,800–2,600	500–750	**700–1,300**	−350–−600	350–550	−850–1,700
Resolution (ppb equivalent)	<4	<10	**<5**	<5	<3	<3
Response time (t90) (s)	<50	<25	**<30**	<35	<50	<30
Zero current (ppb equiv. @ 20 °C)	<−15–+5	<−400–0	**<−100–+500**	<−60–+100	<0–+200	<−75–0
Overgas Limit (ppm)	100	100	**50**	50	50	nd
Operating life (months)[a]	>24	>24	**>24**	>24	>24	>24
Temperature range (°C)	−30–50	−30–50	**−30–50**	−30–50	−30–50	−20–50
Pressure range (kPa)	80–120	80–120	**80–120**	80–120	80–120	80–120
Humidity range (%rh)	15–90	15–90	**15–90**	15–90	15–90	15–90
Load resistor (ohm) recommended	33–100	33–100	**33–100**	33–100	33–100	33–100

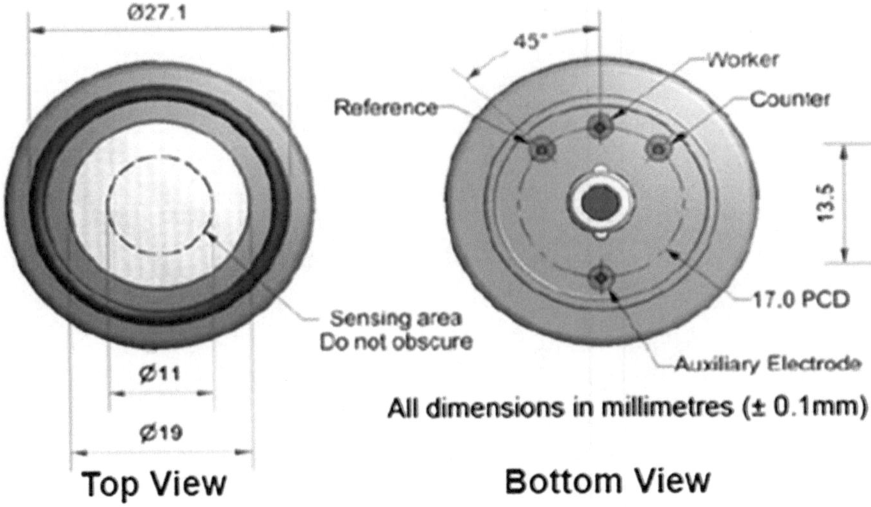

Fig. 8.4 NO-B4 schematic diagram [46]

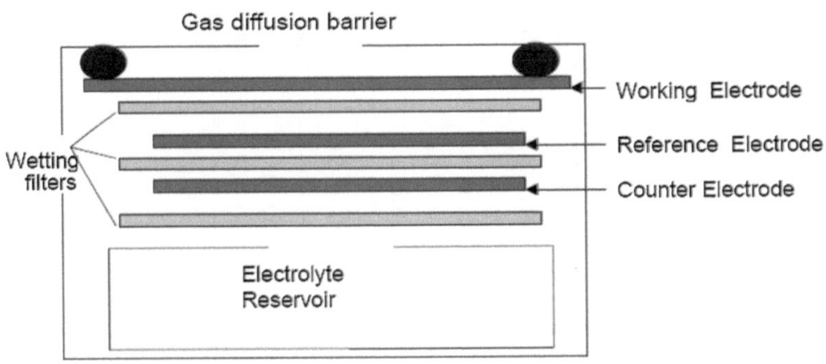

Fig. 8.5 Basic structure of NOB4 [46]

of the circuit diagram and other details of sensor can be obtained from the manufacturer's website [47]. Figure 8.7 shows the sensor schematic and signal conditioning circuit [47].

TGS sensor complies with the required characteristics when the sensor is tested in standard conditions and is specified as follows:

Test Gas Conditions: 20° ± 2 °C, 65 ± 5 %R.H.
Circuit Conditions: $V_C = 15.0 \pm 0.1$ V (DC),
$V_H = 5.0$ V DC ± 5 %,
$R_L =$ Variable $P_s = \leq$ 15mW
Conditioning period before testing: 2–7 days
Sensor Resistance (*Rs*) is calculated by the following formula:

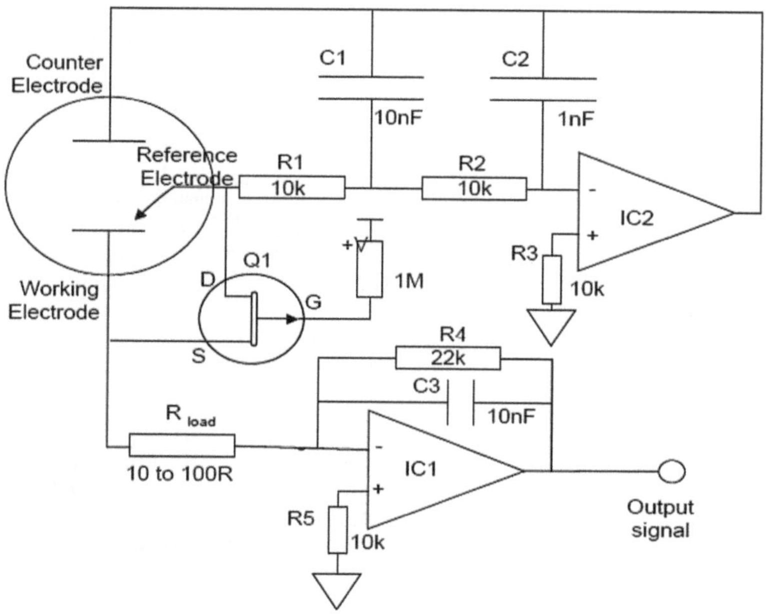

Fig. 8.6 Preferred potentiostat circuit for zero bias toxic gas sensors [46]

$$R_s = (V_c/V_{RL} - 1) \times R_L$$

Power dissipation across sensor electrodes (P_s) is calculated by the following formula: (Table 8.10)

$$P_s = \frac{(V_C \times V_{RL})^2}{R_S}$$

The hardware of the system consists of a microcontroller and some analog components to amplify current. Another part handles the communication with the computer. Gas sensors can be connected to the board. Gas sensors can detect different substances of organic vapor. Photographs of the developed hardware prototype E-nose model are shown in Fig. 8.8.

Software Implementation

Virtual Instrument Software Architecture (VISA) is a standard interface library comprising RS-232 and other protocols. VISA provides the programming interface between the hardware and development environments. LabVIEW comes with interactive tools and configuration utilities for VISA.

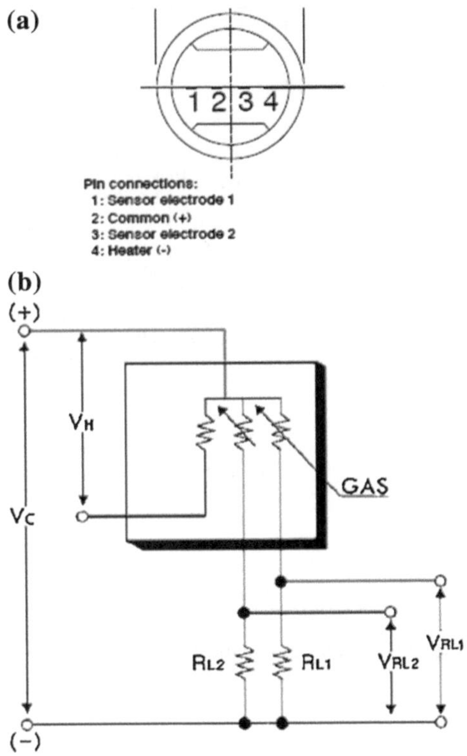

Fig. 8.7 **a** Pin connection of TGS sensor **b** conditioning circuit

The total time duration of the test is of 25 s. The initial 15 s are spent to get a steady-state level of the TGS sensor. While the sensor accommodates with the environment, the timer clearly displays the time elapsed. After the 15th second, the message box displays "Apply Breath" and the test is made for another 10 s. After completion of 25 s, the VI filters the signal using wavelet denoising and calibrates the results using peak and valley detection of the sensor response. The ultimate change is calculated and according to the change measured with sensor, the system gives results of severity of asthma. Figures 8.9 and 8.10 display the patient data entry window and level of asthma detection with prescribed medicines.

8.1.10.3 Result

The sensor used for detection of nitrogen oxide level and the sensor response for different patients corresponding to nitrogen oxide (NO) level in their breath is shown in Figs. 8.11, 8.12, and 8.13: result window. Sensor indicates the presence of NO in terms of corresponding amplitude signal. The amplitude of the sensor output signal varies in accordance with the NO level.

The reference line is formed through the breath sample analysis of healthy people. From the sensor responses of healthy breath samples, the threshold level of

Table 8.10 TGS sensor details: adapted from sensor (www.figarosensor.com)

Gas type	Sensor types	Range of sensor
Combustible gases		
LP-gas/propane	TGS813, TGS2610, TGS6810	(500–10,000 ppm)
Natural gas/methane	TGS842, TGS2611, TGS6810	(500–10,000 ppm)
General combustible gas	TGS813, TGS2610	(500–10,000 ppm)
Hydrogen	TGS821, TGS6812	(50–1,000 ppm)
Toxic gases		
Carbon monoxide	TGS2442, TGS5042	(50–1,000 ppm)
Ammonia	TGS826, TGS2444	(30–300 ppm)
Hydrogen sulfide	TGS825	(5–100 ppm)
Combustible/toxic gas		
Methane	TGS3870	500–12,500 ppm
Carbon monoxide		50–1,000 ppm
Organic solvents		
Alcohol, toluene, xylene & Other volatile organic vapors	TGS822, TGS2620	(50–5,000 ppm)
CFCs (HCFCs and HFCs)- halocarbon		
R-22, R-113	TGS830	(100–3,000 ppm)
R-21-R-22	TGS831	(100–3,000 ppm)
R-134a,R-22	TGS832	(100–3,000 ppm)
Other refrigerant gases	TGS83x series,TGS2630	–
Indoor pollutants		
Carbon dioxide	TGS4161, TGS4160	350–10,000 ppm
Air contaminants	TGS800, TGS2600, TGS2602	(<10 ppm)
Automobile ventilation		
Gasoline exhaust, gasoline and diesel exhaust	TGS2104, TGS2201	0.1–10 ppm(NO) 10–1,000 ppm (CO, H_2,HC)
Cooking vapors		
Volatile vapors from food (alcohol), water vapors from food	TGS880, TGS883T	–

sensor output amplitude is decided, which is marked as reference line. Sensor output amplitude is directly proportional to the intensity of nitrogen oxide (NO) in the breath sample.

8.1.10.4 Conclusion

The portable device that can detect the variety of diseases associated with the human breath can be designed by escalating the present prototype. We want to develop the prototype E-nose which works easily for diagnosis of disease and is

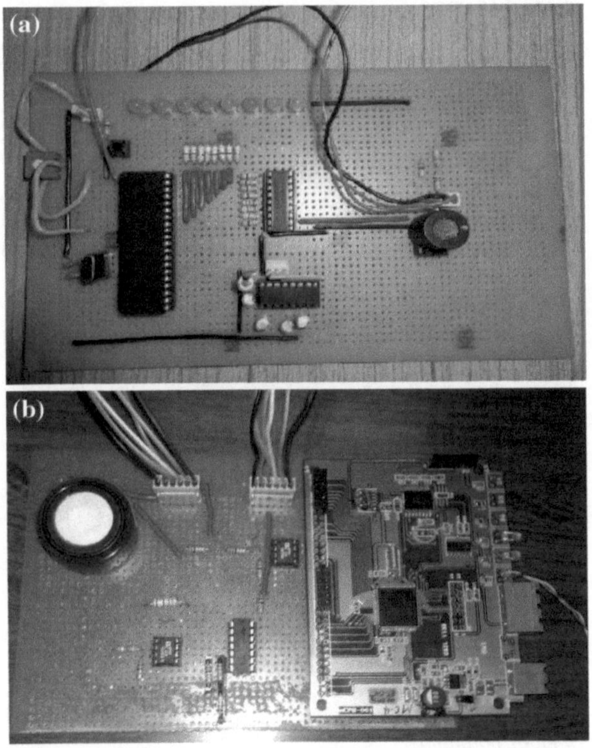

Fig. 8.8 **a** Hardware implementation with the use of sensor (type 1) **b** Sensor (type 2)

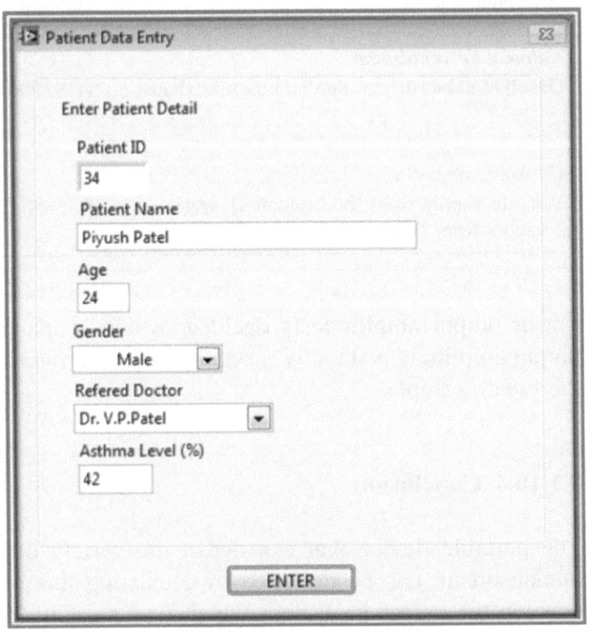

Fig. 8.9 Patient data entry window

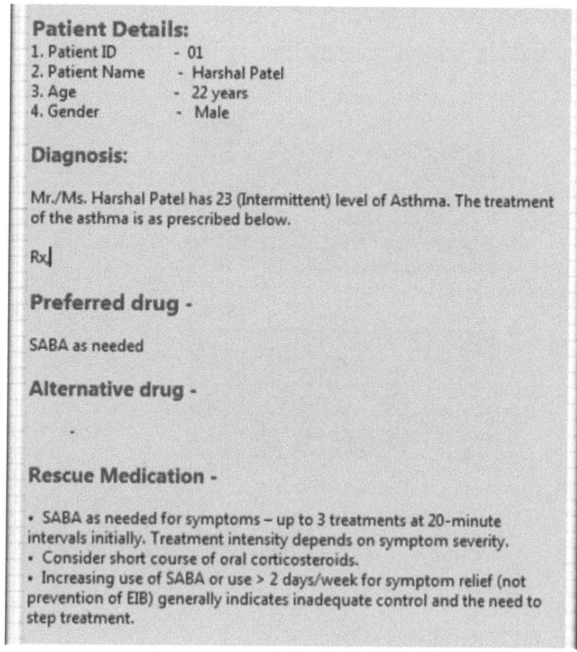

Fig. 8.10 Asthma level with drug doses window

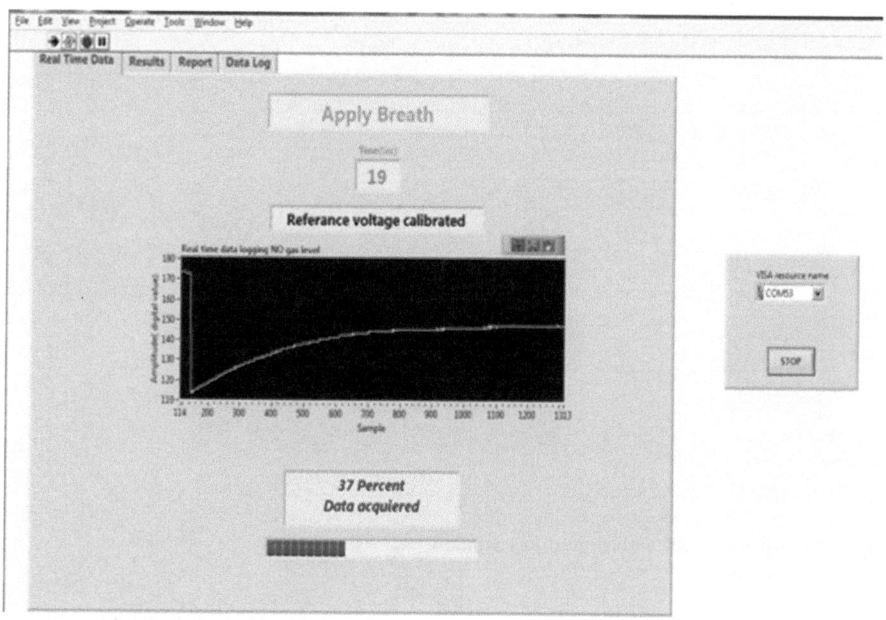

Fig. 8.11 Real-time data logging window

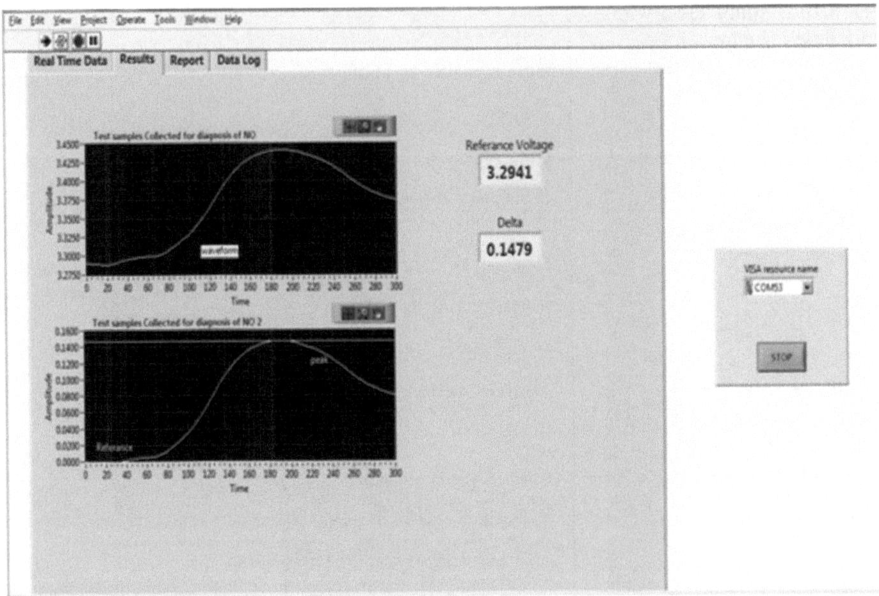

Fig. 8.12 Results window

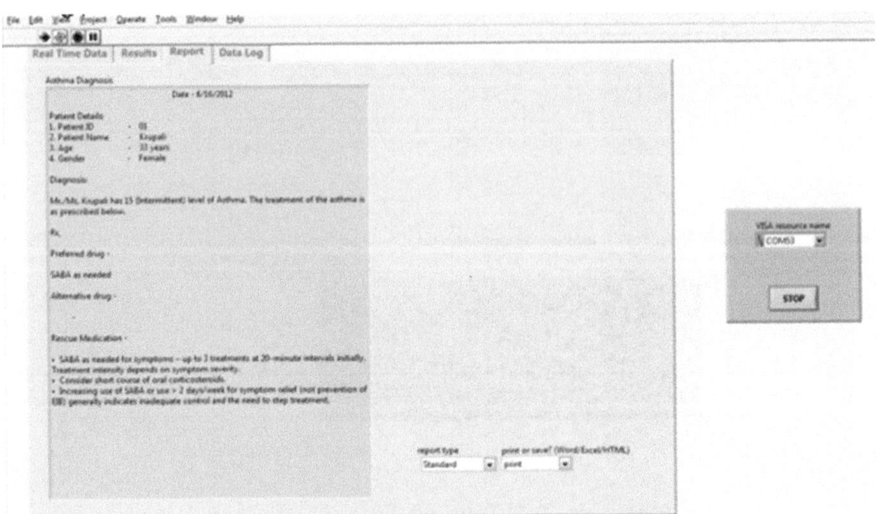

Fig. 8.13 Asthma level with drug doses window

easy to operate for a medical person. Toxic gas sensor sensitivities are variable, so we must calibrate it in the software to correct the sensor-to-sensor sensitivity variations. Sensor sensitivity will also drift downwards with time, depending on the sensor type, relative humidity, and gas concentration/temperature conditions. The sensor signal conditioning test period is long. The main thing is sensitivity of sensor for NO detection which varies as different types of sensors available in the market are the same, so as per application definition the sensor type can be selected and accordingly the price of the sensor varies. Sensor requires a low noise potentiostat circuit for lowest noise and best resolution (i.e., NOB4).

8.2 Case Study 2: Practical Approach to Detect Alcohol in Human Breath

8.2.1 The Problem

Most frequently mentioned observation of traffic police officers in alcohol-related traffic offences is alcohol breath odor. The odor frequency is categorized as slight, moderate, or strong.

An organic compound containing carbon and hydrogen atoms and a hydroxyl (–OH) group is identified as alcohol. Alcohols are infinitely soluble in water, weak acids, and are clear and colorless. Alcohols are of various types and all have their own properties and toxicities and as such death will result if a sufficient quantity is consumed or else introduced into the body. Table 8.11 shows the various types of alcohols and their properties [48].

Detection of alcohol concentration in the brain is essential for safety on the road as well as in the workplace. Suitably, blood alcohol concentration (BAC), defined as the percentage of alcohol in the blood, is used to assess the alcohol level in the brain tissue as a measure of impairment from alcohol poisoning.

Most people show measurable mental mutilation at about 0.05 % BAC. Above this level, the capability to operate an automobile deteriorates gradually with increased blood alcohol level. For the average person, an unconsciousness

Table 8.11 Types of alcohols

Name	Formula	Boiling point (°C)	Uses	Toxicity and metabolites
Methanol methyl alcohol wood alcohol	CH_3OH	64.5	Denaturant solvent paint remover fuel	~75 ml Formaldehyde
Ethanol ethyl alcohol grain alcohol	C_2H_5OH	78.3	Beverage solvent medicinal Veh. fuel	~400–500 ml Acetaldehyde
Isopropanol isopropyl alcohol rubbing alcohol	C_3H_7OH	82.3	Denaturant antiseptic	~250 ml Acetone
Butanol butyl alcohol	C_4H_9OH	117	Perfume and aftershave Base	~100 ml

Table 8.12 Various human symptoms according to the BAC level [49]

Blood alcohol concentration (BAC) (%)	Symptoms
>0.05	Measurable mental impairment
>0.10	Unsteady gait
>0.15	Slurred speech
>0.40	Unconsciousness
>0.50	Difficulty in breathing, heart failure, death

Table 8.13 Breath and blood alcohol concentration unit

Breath alcohol concentration (BRAC) units	
milligrams per liter	mg/l
Micrograms per 100 ml	μg/100 ml, μg %
Microgram per liter	μg/l
Breath alcohol concentration (BAC) units	
Milligrams per 100 ml	Mg/100 ml, mg %
Percent blood alcohol [grams of alcohol per 100 ml of blood or 210 l of breath]	%BAC, %BAL
Promiile [weight/volume] [grams of alcohol per liter of blood]	‰ g/l
Promiile[weight / weight] [grams of alcohol per kilogram of blood]	‰ g/kg

outcome is about a BAC of 0.4 %. Above 0.5 % BAC, basic body functions such as breathing or the beating action of the heart can be low down leading to death. The major symptoms taking place in a person intoxicated by alcohol are listed in Table 8.12 and units of breath and blood alcohol concentration are listed in Table 8.13 [49].

In modern times, the number of traffic accidents due to drunken state of drivers has increased extensively. By checking the concentrations of the gas breathed by a driver, we can evaluate whether a driver drinks alcohol and also know the quantity of the alcohol. Herbert et al. found in their restricted environment study and demonstrated that officers were able to derive only limited information from alcohol odor. Also note that for the low probability of detecting an alcohol breath odor, it might be wise for officers to use a breath testing device whenever a driver reveals behaviors associated with alcohol use [50].

Gases come out in the exhaled breath as by-products of different metabolic pathways or after oral eating/drinking can be detected and measured. Hence, ethanol measurement has set up broad applications in the control of alcohol use by drivers. One interesting device developed by installing devices called Alcolocks or alcohol interlock systems in transportation vehicles in Sweden [8], is an automatic control system designed to prevent driving after excessive alcohol intake by requiring the driver to blow into an in-car breathalyzer before starting the ignition. The alcohol interlock can be set at different levels and limits.

8.2.2 Methods

David Tinwin describes four types of breath alcohol devices which include semiconductor models (Breathalyzer), fuel cell models (alcosensors), infrared (IR) spectroscopy models (intoxilyzers), and Gas Chromatography (GC) models (intoximeters) [51].

Taking blood to check alcohol on the road is difficult and almost impossible, so the simple and feasible method is to check the extent of the gas breathed by a driver. The E-nose can test the alcohol quantity of a driver conveniently and save important data at any time.

8.2.2.1 Sensor

The wide-ranging selectivity of the chemical sensors in an E-nose is remunerated by advanced information processing, although the sensor is a key design parameter for the system. These design parameters include sensitivity, speed of operation, cost, size, manufacturability, ability to operate in diverse environments, and the ability to be automatically and quickly cleaned.

MQ-3 (Hanwei electronics: www.hwsensor.com) gas sensor has good sensitivity to alcohol, and has high quality resistance to disturbance from gasoline, smoke, and vapor. The sensor could be used to detect alcohol at different concentrations; it is low cost and suitable for different applications. Basically, it has six pins, the cover, and the body. Of the six pins, only four can be used. Two are for the heating system, called H, and the other two are for connecting power and ground, called A and B in Fig. 8.14.

Also, TGS -822 (Figaro) can be used as breath alcohol detectors [47]. The Figaro sensor developed using a tin dioxide (SnO_2) semiconductor as sensing element of sensor has low conductivity in clean air. The sensor's conductivity increases depending on the presence of detectable gas concentration in the air. As shown in Fig. 8.13 a simple electrical circuit can convert the change in conductivity into an output signal which corresponds to the gas concentration.

The main characteristic of the sensor (TGS 822) is high sensitivity to the vapors of organic solvents as well as other volatile vapors. Note that it also has sensitivity to a variety of combustible gases such as carbon monoxide, making it a good general-purpose sensor.

It is also available with a ceramic base which is highly resistant to severe environments as high as 200 °C (model# TGS 823) Fig. 8.15.

TGS sensor complies with the above electrical characteristics when the sensor is tested in standard conditions as specified:

Test Gas Conditions: 20° ± 2 °C, 65 ± 5 %R.H.
Circuit Conditions: $V_C = 10.0 \pm 0.1$ V (AC or DC),
$V_H = 5.0 \pm 0.05$ V (AC or DC),
$R_L = 10.0\ k\Omega \pm 1\ \%$

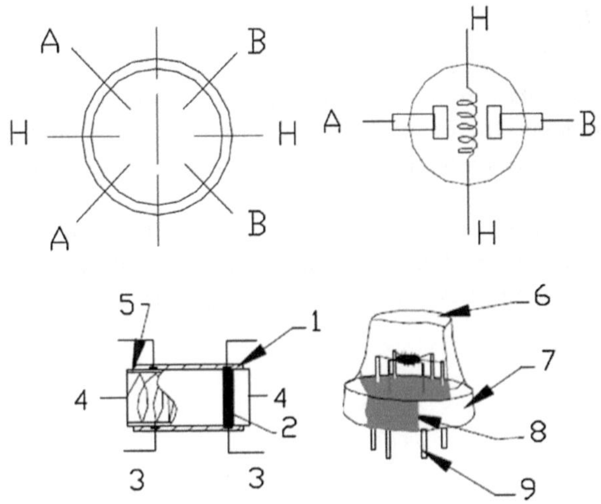

	Parts	Material
1	Gas sensing layer	SnO_2
2	Electrode	Au
3	Electrode line	Pt
4	Heater coil	Ni-Cr alloy
5	Tubular ceramic	Al_2O_3
6	Anti-explosion network	Stainless steel gauze
7	Clamp ring	Copper plating Ni
8	Resin base	Bakelite
9	Tube pin	Copper plating Ni

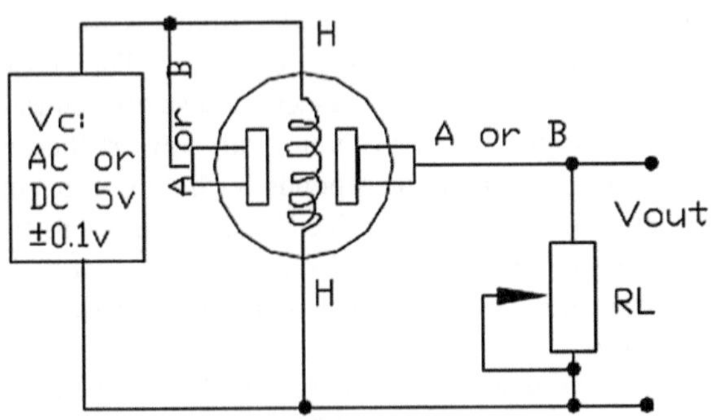

Fig. 8.14 Diagram details of MQ-3 sensor [51]

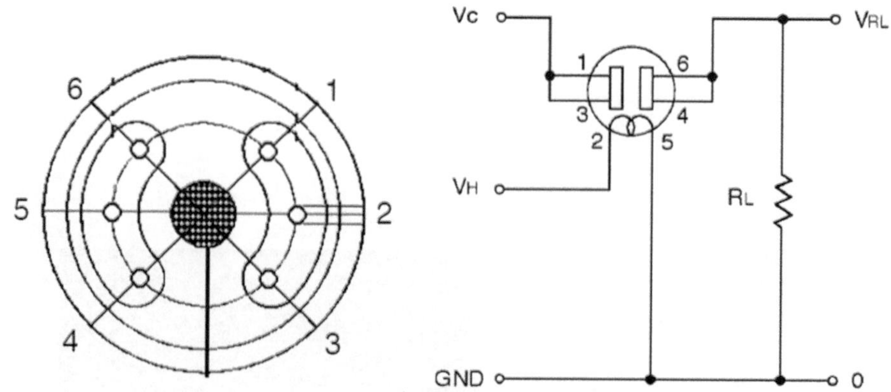

Pin no.1 & 3	V_C – 24V (Max.)
Pin No.2	V_H(Heater Voltage) - 5.0±0.2V
Pin No. 5	Gnd (Circuit Voltage)
Pin No. 4 & 6	V_{RL}(Load resistance) variable 0.45K (Min.)

Fig. 8.15 Alcohol breath sensor [47]

Preheating period before testing: More than 7 days
Sensor Resistance (*Rs*) is calculated by the following formula:

$$R_s = (V_c/V_{\mathbf{RL}} - 1) \times R_L$$

Power dissipation across sensor electrodes (P_s) is calculated by the following formula:

$$P_s = \frac{V_{C^2} \times R_S}{(R_S + R_L)^2}$$

8.2.3 Results

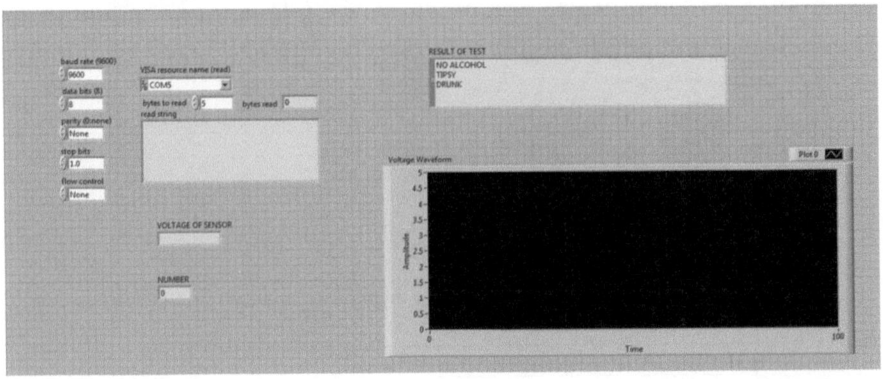

Figure 8.16 shows a measurement with the E-nose measurement system. Figure 8.17 shows real experiments with the E-nose system. The liquid that was tested was pure ethanol. Experiments were conducted and results were taken again and again and compared. The final result generated in LabVIEW is that how much concentration of alcohol is released in the breath of a person and a message is displayed whether the person is “drunk,” “tipsy,” or “no alcohol.”

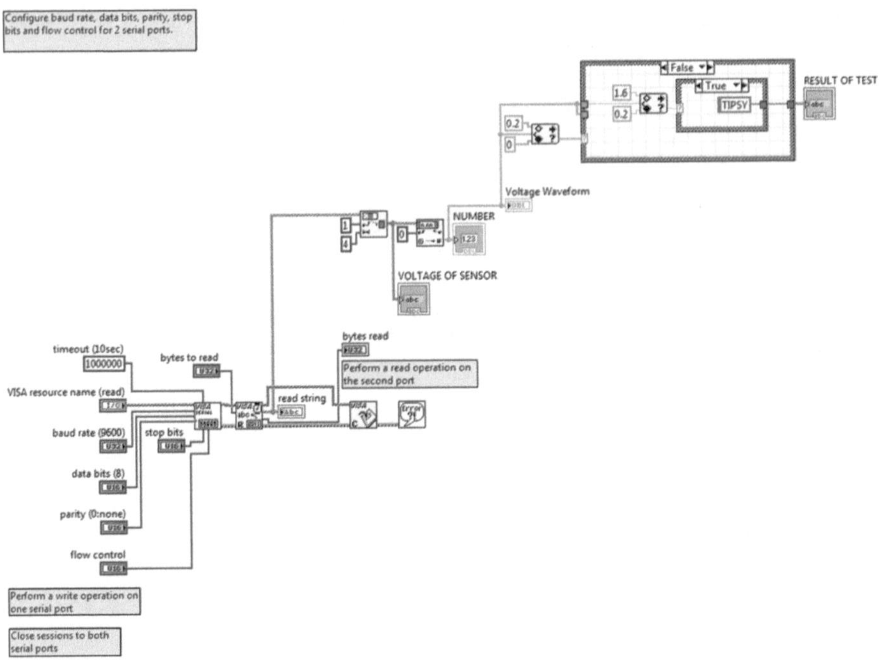

Fig. 8.16 Block diagram in Lab VIEW for serial reception, comparison of values, and display of result window

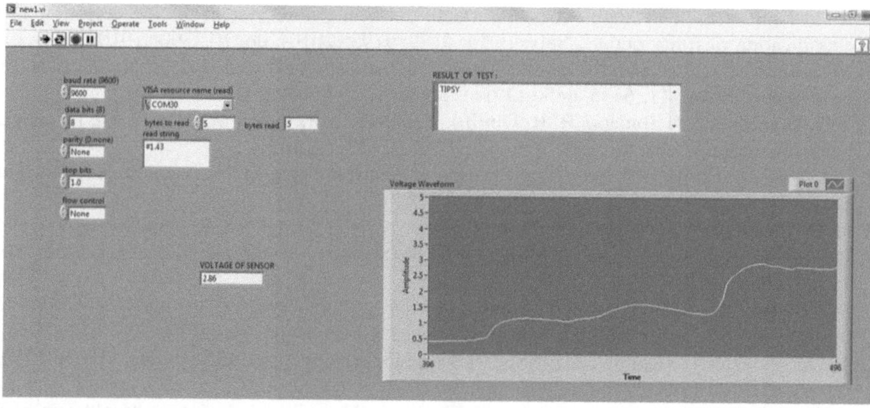

Fig. 8.17 Front panel of Lab VIEW for result display

8.2.4 Conclusion

The E-nose design has a wide area of application. But as the sensors used take a good amount of time and preheating (i.e., TGS Sensor required preheating period before testing for more than 7 days) to give stable output. Also, it was quite difficult to calibrate the sensor (i.e., MQ3 sensor sensitivity is different to various kinds and various concentrations of gases, so sensitivity adjustment is necessary). Finally, after repeated testing of the sensor, with fine calibration of the system, reliable and accurate results can be obtained.

References

1. O. Özgür, B. Karlık, An Overview of Metal Oxide Semiconducting Sensors in Electronic Nose Applications, International Burch University, Department of Information Technology
2. A.D. Wilson et al., Applications and advances in electronic-nose technologies. Sensors **9**, 5099–5148 (2009). doi:10.3390/s90705099
3. P. Russell, Sensory analysis. Milk Ind. Int. **97**, 11–12 (1995)
4. D. Sivalingam, J.B. Balaguru Rayappan, Development of E-nose prototype for raw milk quality discrimination. Milchwissenschaft **67**(4), 381 (2012)
5. S. Ampuero et al., The electronic nose applied to dairy products: A review. Sens. Actuators B **94**, 1–12 (2003)
6. K.M. Horváth, Z.S. Seregely, I. Dalmadi, E. Andrassy, J. Farkas, Estimation of bacteriological spoilage of pork cutlets by electronic nose. Acta Microbiol. Immunol. Hung. **54**(2), 179–194 (2007)
7. M. Ghasemi-Varnamkhasti et al., Meat quality assessment by electronic nose (machine olfaction technology). Sensors **9**, 6058–6083 (2009). doi:10.3390/s90806058
8. H. GholamHosseini, D. Luo, H. Liu, G. Xu, Intelligent processing of E-nose information for fish freshness assessment. 3rd international conference on "intelligent sensors, sensor networks and information", 2007. ISSNIP 2007

9. K.-T. Tang et al., Development of a portable electronic nose system for the detection and classification of fruity odours. Sensors **10**, 9179–9193 (2010). doi:10.3390/s101009179
10. M. Mamat et al., An electronic nose for reliable measurement and correct classification of beverages. Sensors **11**, 6435–6453 (2011). doi:10.3390/s110606435
11. V.O. Olunloyo, T.A. Ibidapo, R. R. Dinrifo, Neural network-based electronic nose for cocoa beans quality assessment. Agric. Eng. Int. CIGR J. **13**(4) (2011)
12. F. ČAČIĆ, L. PRIMORAC et al. Application of electronic nose in honey geographical origin characterization. J. Central Eur. Agric. **10**(1) (2009)
13. N. Bhattacharyya, S. Seth, B. Tudu, P. Tamuly, A. Jana, D. Ghosh, R. Bandyopadhyay, M. Bhuyan, Monitoring of black tea fermentation process using electronic nose. J. Food Eng. **80**, 1146–1156 (2007)
14. S. Linehan, On the application of a consumer preference-based method for designing products to wine fermentation monitoring devices. Chem. Eng. Comm. **198**:255–272 (2011). Taylor & Francis Group, LLC,ISSN: 0098-6445 print = 1563-5201 online, doi:10.1080/00986445.2010.499833
15. A. D. Wilson "Future Applications of Electronic-Nose Technologies in Healthcare and Biomedicine, Wide Spectra of Quality Control", Dr. Isin Akyar (Ed.), ISBN: 978-953-307-683-6, In Tech (2011)
16. M. Trincavelli, S. Coradeschi, A. Loutfi, B. Söderquist, P. Thunberg Member, IEEE, Direct identification of bacteria in blood culture samples using an electronic nose. IEEE Trans. Biomedical Eng. **57**(12), 2884–2890 (2010)
17. S.Y. Lai, O.F. Deffenderfer, W. Hanson, M.P. Phillips, E.R. Thaler, Identification of upper respiratory bacterial pathogens with the electronic nose. Laryngoscope **112**, 975–979 (2002)
18. M. L. Humphreys, R. Orme, N. Sahgal, C. Kendall,N. Magan, N. Stone Electronic nose analysis of bronchoalveolar lavage fluid for the diagnosis of ventilator-associated pneumonia. *Intensive Care Society's (ICS) State of the art meeting*, December (2007), London, UK
19. S. Aathithan, J.C. Plant, A.N. Chaudry, G.L. French, Diagnosis of bacteriuria by detection of volatile organic compounds in urine using an automated headspace analyzer with multiple conducting polymer sensors. J. Clin. Microbiol. **39**, 2590–2593 (2001)
20. A.K. Pavlou, N. Magan, C. McNulty, J. Jones, D. Sharp, J. Brown, A.P. Turner, Use of an electronic nose system for diagnoses of urinary tract infections. Biosens. Bioelectron. **17**, 893–899 (2002)
21. A.K. Pavlou, N. Magan, J.M. Jones, J. Brown, P. Klatser, A.P. Turner, Detection of mycobacterium tuberculosis (TB) in vitro and in situ using an electronic nose in combination with a neural network system. Biosens. Bioelectron. **20**, 538–544 (2004)
22. R. Fend et al., Monitoring haemodialysis using electronic nose and chemometrics. Biosens. Bioelectron. **19**(12), 15 (2004)
23. R. F. Machado, Detection of lung cancer by sensor array analyses of exhaled breath. Am J Respir Crit Care Med **171**: 1286–1291 (2005). doi:10.1164/rccm.200409-1184O on March 4, 2005
24. R. Blatt, A. Bonarini, E. Calabro, M. Della Torre, M. Matteucci, U. Pastorino, Lung cancer identification by an electronic nose based on an array of MOS sensors. Neural Networks, IJCNN (2007), pp. 1423–1428
25. D. Guo, D. Zhang, N. Li, L. Zhang, J. Yang, A novel breath analysis system based on electronic olfaction. IEEE Trans. Biomed. Eng. **57**(11), 2753–2763 (2010)
26. Arend Kolk et al., Electronic-nose technology in diagnosis of TB patients using sputum samples. J. Clin. Microbiol. (2010). doi:10.1128/JCM.00569-10
27. N. Charaklias, H. Raja, M.L. Humphreys, N. Magan, C.A. Kendall, The future of early disease detection: Applications of E-nose technology in otolaryngology. J. Laryngol. Otol. **124**(8), 823–827 (2010)
28. M. T. Momol, M. O. Balaban, F. Korel, A. Odabasia, E. A. Momel, G. Folkes, J. B. Jones, Discrimination of plant pathogenic bacteria using an electronic nose, Online. Plant health Progress, (2004)

29. A.C. Bastos, N. Magan, Soil volatile fingerprints: Use for discrimination between soil types under different environmental conditions. Sens. Actuators B **125**, 556–562 (2007)
30. Alphus D. Wilson, Diverse applications of electronic-nose technologies in agriculture and forestry. Sensors **13**, 2295–2348 (2013). doi:10.3390/s130202295
31. A. Catarina Bastos, N. Magan, Potential of an electronic nose for the early detection and differentiation between Streptomyces in potable water. Sens Actuators B **116**, 151–155 (2006)
32. W. Bourgeois, R.M. Stuetz, Measuring wastewater quality using a sensor array: prospects for real-time monitoring. Water Sci. Tech. **41**(12), 107–112 (2000)
33. S. Zampolli et al., An electronic nose based on solid state sensor arrays for low-cost indoor air quality monitoring applications. Sens. Actuators B **101**, 39–46 (2004)
34. F.D. Francesco et al., An electronic nose for odour annoyance assessment. Atmos. Environ. **35**, 1225–1234 (2001)
35. K. C. Persaud, P. Wareham, A. M. Pisanelli, Emmanuel scorsone, 'electronic nose'- new condition monitoring devices for environmental applications. Chem. Senses **30** (suppl 1): i252–i253 (2005)
36. J.E. Staples, *The First Quantitatively Validated Electronic Nose for Environmental Testing of Air, Water, and Soil* (ACS National, March, 2000), pp. 26–30
37. M. Bonnefille et. al. Prospective experiments of E-nose for cosmetic applications: recognition of sweat odours, agro-industrial chemistry laboratory, France
38. G.W. Watson, D.S. McGuire, *Detection of explosives in soil and water with an Electronic Nose* (American Chemical Society Meeting, Ontario, California, 1999), pp. 5–7. October
39. M.C. Burl et al., Mining the detector responses of a conducting polymer composite-based electronic nose. First SIAM Int. Conference on Data Mining, (2000)
40. M. A. Ryan et. al., Monitoring space shuttle air quality using the jet propulsion laboratory electronic nose. IEEE Sens J **4**(3) (2004)
41. L. Zhu et al., Flavor analysis in a pharmaceutical oral solution formulation using an electronic-nose. J. Pharm. Biomed. Anal. **34**, 453–461 (2004)
42. D.H. Yates, Role of exhaled nitric oxide in asthma. Immunol. Cell Biol. **79**(2), 178–190 (2001)
43. K. Alving, E. Weitzberg, J.M. Lundberg, Increased amount of nitric oxide in exhaled air of asthmatics. Eur. Respir. J. **6**(9), 1368–1370 (1993)
44. W.Q. Cao, Y.X. Duan, Breath analysis: Potential for clinical diagnosis and exposure assessment. Clin. Chem. **52**(5), 800–811 (2006)
45. L.J. Dupont, M.G. Demedts, G.M. Verleden, Prospective evaluation of the validity of exhaled nitric oxide for the diagnosis of asthma. Chest **123**(3), 751–756 (2003)
46. www.alphasense.com
47. http://www.figaro.co.jp/
48. Department of forensic science, "Breath test operator instructional manual", (2005)
49. NBS Special Publication 480–41
50. H. Moskowitz et al., Police officers' detection of breath odours from alcohol ingestion. Accid. Anal. Prev. **31**, 175–180 (1999)
51. D. Tinwin, Breath alcohol testers: prevents road accidents. AU J.T. **10**(2): 75–80 (2006)
52. Dr Gambert, Breath analysis with electrochemical sensors, GmBH, Germany. http://www. it-wismar.de, IGAMED-workshop, (2007)
53. L. Wang, *Tailored synthesis and characterization of selective metabolite-detecting nanoprobes for handheld breath analysis* (Stony Brook University, December, Dissertation, 2008)
54. B. Hök, H. Pettersson, A.K. Andersson, S. Haasl, P. Åkerlund, Breath analyzer for alcolocks and screening devices. IEEE Sens. J. **10**, 10–15 (2010)
55. http://www.hwsensor.com (HANWEI Electronics Co., Ltd,)

About the Author

Prof. H. K. Patel holds a Bachelor of Engineering degree in Instrumentation and Control Engineering and a Master of Technology degree in Electrical Engineering. His major areas of expertise are Sensors, Measurement Systems, Industrial Electronics and Biomedical Engineering. He is currently an Associate Professor in Instrumentation and Control Engineering at Electrical engineering department of Nirma University, Ahmedabad, India. At present, he is pursuing his Ph.D. and his major area of research is focused at the development of a prototype electronic nose system for human disease diagnosis via breath sampling.

He is an active member of several International as well as national professional organizations, including the Institute of electrical and electronics engineers (IEEE) and the IEEE's instrument and computer societies and instrument society of America (ISA). He is also a life member of Indian society for technical education (ISTE).

Prof. Patel has received an IEE national award for "The Best Technical Paper" in 2005. He has published several other book titles and more than 40 research papers in reputed international journals and international conference proceedings. He has served as session chair in many international conferences and has been an esteemed reviewer for many international journals.

H. K. Patel, *The Electronic Nose: Artificial Olfaction Technology*,
Biological and Medical Physics, Biomedical Engineering,
DOI: 10.1007/978-81-322-1548-6, © Springer India 2014

About the Author

Index

H. K. Patel, *The Electronic Nose: Artificial Olfaction Technology*, Biological and Medical Physics, Biomedical Engineering, DOI: 10.1007/978-81-322-1548-6, © Springer India 2014